DRAWING
LOUISIANA'S NEW MAP

ADDRESSING LAND LOSS IN
COASTAL LOUISIANA

Committee on the Restoration and Protection of Coastal Louisiana

Ocean Studies Board

Division on Earth and Life Studies

NATIONAL RESEARCH COUNCIL
OF THE NATIONAL ACADEMIES

THE NATIONAL ACADEMIES PRESS
Washington, D.C.
www.nap.edu

THE NATIONAL ACADEMIES PRESS   500 Fifth Street, N.W.   Washington, DC 20001

NOTICE: The project that is the subject of this report was approved by the Governing Board of the National Research Council, whose members are drawn from the councils of the National Academy of Sciences, the National Academy of Engineering, and the Institute of Medicine. The members of the committee responsible for the report were chosen for their special competences and with regard for appropriate balance.

This study was supported by Cooperative Agreement No. 2512-03-01 and Cooperative Agreement No. 435-300458 between the National Academy of Sciences and the State of Louisiana with support from the U.S. Army Corps of Engineers. Any opinions, findings, conclusions, or recommendations expressed in this publication are those of the author(s) and do not necessarily reflect the views of the organizations or agencies that provided support for the project.

International Standard Book Number 0-309-10054-2

Library of Congress Catalog Card Number 2006920074

Cover: Images of the cypress swamp and navigation canal were provided by Dr. Donald Davis of Louisiana State University.

Additional copies of this report are available from the National Academies Press, 500 Fifth Street, N.W., Lockbox 285, Washington, DC 20055; (800) 624-6242 or (202) 334-3313 (in the Washington metropolitan area); Internet, http://www.nap.edu.

THE NATIONAL ACADEMIES
Advisers to the Nation on Science, Engineering, and Medicine

The **National Academy of Sciences** is a private, nonprofit, self-perpetuating society of distinguished scholars engaged in scientific and engineering research, dedicated to the furtherance of science and technology and to their use for the general welfare. Upon the authority of the charter granted to it by the Congress in 1863, the Academy has a mandate that requires it to advise the federal government on scientific and technical matters. Dr. Ralph J. Cicerone is president of the National Academy of Sciences.

The **National Academy of Engineering** was established in 1964, under the charter of the National Academy of Sciences, as a parallel organization of outstanding engineers. It is autonomous in its administration and in the selection of its members, sharing with the National Academy of Sciences the responsibility for advising the federal government. The National Academy of Engineering also sponsors engineering programs aimed at meeting national needs, encourages education and research, and recognizes the superior achievements of engineers. Dr. Wm. A. Wulf is president of the National Academy of Engineering.

The **Institute of Medicine** was established in 1970 by the National Academy of Sciences to secure the services of eminent members of appropriate professions in the examination of policy matters pertaining to the health of the public. The Institute acts under the responsibility given to the National Academy of Sciences by its congressional charter to be an adviser to the federal government and, upon its own initiative, to identify issues of medical care, research, and education. Dr. Harvey V. Fineberg is president of the Institute of Medicine.

The **National Research Council** was organized by the National Academy of Sciences in 1916 to associate the broad community of science and technology with the Academy's purposes of furthering knowledge and advising the federal government. Functioning in accordance with general policies determined by the Academy, the Council has become the principal operating agency of both the National Academy of Sciences and the National Academy of Engineering in providing services to the government, the public, and the scientific and engineering communities. The Council is administered jointly by both Academies and the Institute of Medicine. Dr. Ralph J. Cicerone and Dr. Wm. A. Wulf are chair and vice chair, respectively, of the National Research Council.

www.national-academies.org

OCEAN STUDIES BOARD

Staff

SUSAN ROBERTS, Director
DAN WALKER, Scholar
CHRISTINE BLACKBURN, Program Officer
ANDREAS SOHRE, Financial Associate
SHIREL SMITH, Administrative Coordinator
JODI BOSTROM, Research Associate
NANCY CAPUTO, Research Associate
SARAH CAPOTE, Senior Program Assistant
PHILLIP LONG, Program Assistant

Preface

The first Europeans to arrive in the Louisiana coastal area found an environment of fertile lands rich in natural resources, including fish, shrimp, and fur-bearing animals. Alterations, including construction of levees and constraining the Mississippi River, rendered this system even more agriculturally productive, suitable for habitation, and efficient for navigation. While much of the habitation developed on natural levees, the lowlands yielded bountiful crops of sugar and other valuable agricultural products. Navigation to the nation's heartland through the vast Mississippi River and its tributaries contributed further to the economic vitality of this region. The discovery of petroleum in 1901 provided additional economic stimulus to the region. Today, the harvest of seafood and the unique mix of cultures fuel a significant tourism industry.

Ironically, many of the alterations and activities carried out to increase the general productivity and attractiveness of this region and to exploit its natural resources have contributed to the rapid wetland losses that are the subject of this study. Levees have reduced the deposition of nutrient-rich sediments on areas now in agricultural production, and jetties direct the sediments into deep water that once nourished marshes and barrier islands. Canals cut through marshes for petroleum exploration and for access to production facilities have led to continued degradation of the wetlands, and extraction of hydrocarbons is believed to have augmented the natural subsidence rates. Introduction of nutria has resulted in increased wetland grazing. Finally, constraining the Mississippi River to not allow its natural switching has resulted in a system that is more desirable for navigation but one counter to the natural cycle required for maintaining the net wetland area.

The State of Louisiana and the U.S. Army Corps of Engineers have been aware of the high rates of wetland loss for many years; both have been active in efforts to arrest this loss. In addition to scientific investigations to understand the problem, concerted political efforts commenced in the 1960s that culminated in the passage of the Coastal Wetlands Planning, Protection, and Restoration Act and the further development of plans to arrest or reduce high wetland losses. The present study focused on a near-term plan, a 10-year, $1.9 billion, scaled-down version of the more comprehensive plan (30-year, $13 billion). Broadly, the committee's charge was to assess the economic, engineering, ecological, and social viability of the near-term effort and its value to the nation. This near-term plan is unique in several respects: (1) the vast geographic extent of the coastal Louisiana area (31,080 km^2 [12,000 mi^2]), (2) the pervasiveness of the processes affecting wetland loss, and (3) the amount of background material that has been developed pertinent to an understanding of the wetland loss problem.

The committee and I are indebted to the staff of the Ocean Studies Board for their valuable services and willingness to fill any need ranging from arrangements for the meetings to report editorial services to obtaining additional needed background material. They truly made the efforts of the committee members more enjoyable and productive. Dr. Dan Walker served as program manager for the latter half of the study after Dr. Joanne Bintz departed for other employment. Douglas George was a great help during his tenure as a National Research Council intern with the Ocean Studies Board. We are especially appreciative of Ms. Jodi Bostrom who provided day-to-day support to the committee and, through her broad talents and can-do attitude, ensured that the committee activities were responded to promptly and that unforeseen needs at committee meetings were met. Without the editorial efforts of the staff, this report would be redundant and lack overall flow from topic to topic.

As a final note, the tragedy wreaked on coastal Louisiana by Hurricanes Katrina and Rita occurred after this committee had completed their meetings and after development of a final draft report. Certainly, if this sequence had been different, this report would have had a somewhat different focus. Subsequent to these hurricanes, the committee revisited the thrust of this report relating to our statement of task and recommendations pertaining to wetlands, and it was determined that these previously developed recommendations still applied. It is the hope of the committee that this report will contribute to an understanding of the overall benefits of coastal wetlands and associated levees and barrier islands (not just storm protection) and the urgency of addressing the rapid land loss in coastal Louisiana.

Robert Dean, *Committee Chair*

Acknowledgments

This report was greatly enhanced by participants at the four public meetings held as part of this study. The committee would like first to acknowledge the efforts of those who gave presentations at these meetings: Jack C. Caldwell; Thomas Campbell; Ellis J. (Buddy) Clairain, Jr.; James Coleman; Troy Constance; James Cowan, Jr.; Mark Davis; John Day, Jr.; Gerry Duszynski; Sherwood Gagliano; Karen Gautreaux; Bill Good; James (Randy) Hanchey; Jimmy Johnston; Richard Kesel; Irv Mendelssohn; King Milling; Robert Morton; Shea Penland; Jon Porthouse; Denise Reed; John Saia; Greg Steyer; Greg Stone; Joseph Suhayda; and Robert Twilley. These talks helped set the stage for fruitful discussions in the closed sessions that followed.

The committee is also grateful to a number of people who provided important discussion and/or material for this report: Len Bahr, Windell Curole, Oliver Harmar, Harry Roberts, Robin Rorick, Mark Schleifstein, Kerry St. Pé, Bill Streever, Colin Thorne, and Jeff Williams.

This report has been reviewed in draft form by individuals chosen for their diverse perspectives and technical expertise, in accordance with procedures approved by the National Research Council's Report Review Committee. The purpose of this independent review is to provide candid and critical comments that will assist the institution in making its published report as sound as possible and to ensure that the report meets institutional standards for objectivity, evidence, and responsiveness to the study charge. The review comments and draft manuscript remain confidential to protect the integrity of the deliberative process. We wish to thank the following individuals for their review of this report:

LINDA K. BLUM, University of Virginia, Charlottesville
DONALD F. BOESCH, University of Maryland, Cambridge
LEON E. BORGMAN, University of Wyoming, Laramie
VIRGINIA R. BURKETT, U.S. Geological Survey, Many, Louisiana
CHERYL K. CONTANT, Georgia Institute of Technology, Atlanta
ROBERT FROSCH, Harvard University, Cambridge, Massachusetts
ELVIN R. HEIBERG III, Heiberg Associates, Inc., Arlington, Virginia
PORTER HOAGLAND III, Woods Hole Oceanographic Institution,
    Woods Hole, Massachusetts
WILLIAM F. MARCUSON III, U.S. Army Corps of Engineers (retired),
    Vicksburg, Mississippi
NORMAN H. SLEEP, Stanford University, Stanford, California
PETER R. WILCOCK, Johns Hopkins University, Baltimore, Maryland

Although the reviewers listed above have provided many constructive comments and suggestions, they were not asked to endorse the conclusions or recommendations nor did they see the final draft of the report before its release. The review of this report was overseen by **Frank Stillinger**, Princeton University, and **Gregory Baecher**, University of Maryland. Appointed by the National Research Council, they were responsible for making certain that an independent examination of this report was carried out in accordance with institutional procedures and that all review comments were carefully considered. Responsibility for the final content of this report rests entirely with the authoring committee and the institution.

Contents

SUMMARY   1

1 INTRODUCTION   13
   History and Causes of Land Loss in Louisiana, 14
   History of Coastal Protection in Louisiana, 17
   LCA Study, 23
   Origin and Scope of the Current Study, 25

2 THE HISTORIC AND EXISTING LOUISIANA
   COASTAL SYSTEMS   29
   The Modern, Anthropogenically Modified River and Delta, 32
   The Future Louisiana Coastal System, 42

3 CONFLICTS AND LIMITATIONS TO ACHIEVING GOALS   43
   Land Loss Patterns and Proposed Sediment Distribution, 44
   Stakeholders with Conflicting Interests, 46
   Increasing the Success of the LCA Study's Implementation, 59

4 PLANS AND EFFORTS AT RESTORING COASTAL
   LOUISIANA   63
   Coastal Wetlands Planning, Protection, and Restoration Act, 64
   Coast 2050, 69
   Reconnaissance-Level Report, 72

Draft LCA Comprehensive Study, 73
LCA Study, 79
Implementation of the LCA Study: Organization, Duration, and Funding, 80
Relationship of Coast 2050 and the LCA Study to CWPPRA Projects and Experience, 84
Improving Ongoing Restoration Efforts, 85

5 THE LCA STUDY PLANNING APPROACH, MODELING, AND PROJECT SELECTION PROCESS   87
Context for Planning, 88
Role of Models in the Planning and Adaptive Management of the LCA Study Planning Process, 91
Project Selection and the Link with Modeling, 101
The Improved Modeling and Project Selection Process, 111

6 THE LCA STUDY AND THE FEASIBILITY OF ITS COMPONENTS   115
The Five Major Restoration Features, 118
Other Elements of the LCA Study, 120
Adaptive Management, 123
Proposed Management Approaches, 125
Feasibility, 126
Some Considerations for Long-Term Projects, 133
Enhancing the Feasibility of the Overall Approach, 138

7 CRITICAL KNOWLEDGE GAPS   145
Wetland Loss Causal Factors and Rates, 146
Engineering Knowledge Gaps, 149
Hydrologic Knowledge Gaps, 151
Wetland Formation Knowledge Gaps, 154
Societal Knowledge Gaps, 154
Ecological Knowledge Gaps, 156
Addressing Gaps in the Existing Knowledge Base, 158

8 FINDINGS AND RECOMMENDATIONS   161
Soundness of Approach and Performance Metrics, 162
Addressing Knowledge Gaps, 166
Understanding Costs and Benefits, 169
Economic Justification, 170
Developing a Comprehensive Plan, 172

REFERENCES   177

APPENDIXES
A Committee and Staff Biographies   185
B Acronyms and Abbreviations   189

Summary

On August 29, 2005, Hurricane Katrina struck eastern Louisiana, Mississippi, and western Alabama killing hundreds of local residents, displacing hundreds of thousands more, and causing an estimated $200 billion in economic damage. Less than four weeks later, Hurricane Rita struck easternmost Texas and western Louisiana. Although the loss of life from Rita was much less than that from Katrina, significant destruction resulted, including the reflooding of some parts of New Orleans damaged during Katrina. The devastation wreaked by Katrina and Rita tragically demonstrated the risks that many coastal areas face from hurricanes and associated flooding. Prior to the storms, the historical and continuing land losses in the coastal regions of Louisiana contributed to widespread concerns regarding the vulnerability of the region to storms and coastal flooding. This report, which focuses on restoration efforts proposed by the U.S. Army Corps of Engineers (USACE) and the State of Louisiana in late 2004, was in the final stages of peer review and was essentially complete when the hurricanes struck. The role that land loss, lack of levee maintenance and improvement, and the large navigation channels (including the Mississippi River Gulf Outlet [MRGO]) played in determining the extent of the damage caused by the hurricanes, and their impact on the national economy, cannot be fully determined at this time. What have become clear are the enormous personal, social, economic, and cultural losses that a major hurricane can bring to the residents of the Gulf Coast and the reverberations of such events nationwide. To the extent that wetlands can offset a significant degree of storm impact, large-scale wetlands restoration projects can be an important component of national efforts to reduce future hazards from hurricanes. As more information becomes available about Katrina and Rita, or as reconstruction efforts supersede planned restoration efforts reviewed as part of this study, some of the conclusions of this report may become more significant while others may become moot.

The National Research Council (NRC) recognizes that this report is being released at a time when there may be many more questions than answers. Even so, the report is provided at this difficult time in the hope that its advice on restoring and protecting coastal Louisiana can be considered as part of the nation's strategy to rebuild the Gulf Coast and reduce the likelihood of future tragedies associated with hurricanes in the region.

EFFORTS TO RESTORE AND PROTECT COASTAL LOUISIANA

Coastal wetlands develop within a fine balance of many geomorphologic and coastal ocean processes. Relative sea level rise, wave action, tidal exchange, river discharges, hurricanes and coastal storms, and the rates of sediment accretion due to sediment deposition and accumulation of organic material play particularly important roles. The interplay of these processes and the wetland's resilience to natural or anthropogenic perturbations determine its sustainability. Some of the processes of land loss and gain in the Louisiana coastal area are natural and have occurred for centuries. Others are the result of human activities in the wetlands and the watershed of the Mississippi River system.

Annual land loss rates in coastal Louisiana have varied over the last 50 years, declining from a maximum of 100 square kilometers (km^2) per yr (39 square miles [mi^2] per yr) for the period 1956–1978. Cumulative loss during this 50-year period in Louisiana represents 80 percent of the coastal land loss in the entire United States. Initial efforts to prevent catastrophic land loss were implemented under the federal Coastal Wetlands Planning, Protection, and Restoration Act (CWPPRA) in partnership with Louisiana's efforts through Act 6 (L.A.R.S. 49:213 *et seq.*). Passed in 1990, CWPPRA called for the development of a comprehensive Louisiana Coastal Wetlands Restoration Plan (P.L. 101-646 §303.b). The first such plan was completed in 1993 and has been in use since that time. In addition, the Louisiana Coastal Wetlands Conservation and Restoration Task Force and the Wetlands Conservation and Restoration Authority prepared a plan for the coast in 1998 entitled *Coast 2050: Toward a Sustainable Coastal Louisiana* (Coast 2050).

Coast 2050 was developed under a number of federal and state legislative mandates and is the result of recognition by federal, state, and local agencies that a single plan and coordinated strategy were needed. Coast 2050 was then appended to the 1999 U.S. Army Corps of Engineers 905(b) reconnaissance report. In October 2003, a draft comprehensive study (*Louisiana Coastal Area, LA—Ecosystem Restoration: Comprehensive Coastwide Ecosystem Restoration Study* [draft LCA Comprehensive Study]) for implementing coastal restoration was released. After reviewing the draft LCA Comprehensive Study, the U.S. Office of Management and Budget requested a near-term approach to focus the scope of work and maintain restoration momentum. The resulting final version of *Louisiana Coastal Area (LCA), Louisiana—Ecosystem Restoration Study* (LCA Study) was released by USACE in November 2004. As plans for completion of the LCA Study were being finalized, Louisiana's Office of the Governor requested that the National Academies review the LCA Study's effectiveness for long-term, comprehensive restoration development and implementation.

THE CURRENT STUDY

The LCA Study and its envisioned successors are unique in many respects, including geographic scope, pervasiveness of the destructive processes involved, complexity of potential impacts to stakeholders, success of preceding efforts to achieve stakeholder consensus, and documentation of earlier planning and restoration efforts. Indeed, the environmental and social challenges confronting coastal Louisiana in the near and distant future are without precedent in North America. Clearly, execution of the LCA Study alone will not achieve its stated goal "to reverse the current trend of degradation of the coastal ecosystem," although successful completion of some of the projects outlined in the LCA Study will reduce this trend, thereby representing an important step toward the goal of sustaining or expanding wetlands in some local areas. By definition, the activities proposed in the LCA Study were intended to provide a foundation for successful future restoration and protection efforts, including those developed and implemented in response to hurricanes like Katrina and Rita.

The overall approach taken by the NRC's Committee on the Restoration and Protection of Coastal Louisiana was to examine the LCA Study and all of its components in detail. This examination, supplemented by presentations from key managers, scientists, and engineers in a series of public meetings in Louisiana, served as the basis for evaluating the usefulness of the LCA Study for developing and implementing a long-term comprehensive program consistent with the broad vision articulated in Coast 2050. The committee was charged with addressing four specific sets of questions (shown in italics).

Q: *Are the strategies outlined in the LCA Study based on sound scientific and engineering analyses, and are they appropriate to achieve the goals articulated in the plan? What other approaches might be considered? Are adequate measures of success articulated in the LCA Study?*

Taken individually, the majority of the projects proposed in the LCA Study are based on commonly accepted, sound scientific and engineering analyses. It is not clear, however, that in the aggregate, whether or not these projects represent a scientifically sound strategy for addressing coastal erosion at the scale of the affected area. Thus, at foreseeable rates of land loss, the level of effort described by the LCA Study will likely decrease land loss only in areas adjacent to the specific proposed projects. As stated in numerous USACE policy statements and recommended in past NRC reports, planning and implementation of water resources projects (including those involving environmental restoration) should be undertaken within the context of the larger system. A group of projects within a given watershed or coastal system may interact at a variety of scales to produce either beneficial or deleterious effects. Cost-effectiveness analyses discussed in the LCA Study and in supporting documents reflect an effort to identify least-cost alternatives but do not appear to reflect a system-wide effort to maximize beneficial synergies among various projects. **The selection of any suite of individual projects in future efforts to restore coastal Louisiana should include a clear effort to maximize the beneficial, synergistic effects of individual projects to minimize or reverse future land loss.** Further, because there is a finite availability of water flow and sediment and many of the proposed projects must function for decades to deliver maximum benefit, care should be taken to ensure that implementation of an individual project does not preclude other strategies or elements that are being considered for the future. **To achieve this, the development of an explicit map of the expected future landscape of coastal Louisiana should be a priority as the implementation of the LCA Study moves ahead.**

The approaches advanced in the LCA Study focus largely on proven engineering and other methods to address land loss at the local scale. In general, individual projects appear to be based on commonly accepted, sound scientific and engineering analyses. The emplacement of 61 kilometers [km]) (38 miles [mi]) of revetment along the banks of MRGO as one of the five major wetland restoration projects proposed in the LCA Study, however, does not appear to be consistent with the study's stated goals. Despite an estimated cost of $108.3 million,[1] this project is expected

---

[1]USACE, in the 2005 Chief's Report, updated the cost of the proposed MRGO feature to be $105.3 million.

to reduce land loss by only 0.5 km^2 per yr (0.2 mi^2 per yr) over the next 50 years. (Louisiana is projected to lose an average of 26.7 km^2 per yr [10.3 mi^2 per yr] over the next 50 years.) Although the location of the land loss may make it more significant, the need for and potential value of this project are directly related to the outcome of a study being conducted by USACE, scheduled for completion in FY 2005, to evaluate the potential decommissioning of MRGO for deep draft navigation. In addition to questions regarding the appropriateness of this particular project, its selection casts doubt on the rigor of the ranking and selection process. **The selection of the restoration efforts of MRGO as one of the five major projects to be carried out as part of the LCA Study should be reconsidered in light of the limitations of expected benefits and the results of ongoing studies on the decommissioning of MRGO for deep draft navigation.** If a decision is made to decommission MRGO, various options could be considered, including complete closure, that would significantly reduce the need to strengthen the levees along its route. If partial closure is chosen, perhaps maintaining MRGO for shallow draft vessels, some of the work along the outlet may still be required. Restoration efforts requiring planning would be more fully informed once a final decision has been made.

Conflicting stakeholder interests represent one of the greatest barriers to robust coastal restoration efforts in Louisiana. A dominant human-related component of land loss is the constraint on the river system imposed by spoil banks and levees, but these features also provide benefits to a range of stakeholders. By minimizing the cost of dredging and reducing uncontrolled flooding in inhabited and agricultural areas, these features support important local economic activities. Many of Louisiana's inhabited areas are located on natural levees formed by deposition on the floodplain during major floods. Valuable agricultural land was originally maintained at an elevation above water level through flood-derived sedimentation but is now protected by levees, which preclude new sediment introduction. Obviously, the prospects are low that sediment-rich water will be intentionally allowed to flood broad expanses of urban and agricultural land to maintain elevation with the pace of relative sea level rise.

As discussed above, locating individual projects in an effort to maximize positive synergistic effects will tend to concentrate efforts into selected areas within coastal Louisiana. Although distributing individual projects, and the benefits associated with them, across the entire region may be less contentious, such an approach will either drive up the total cost or reduce the likelihood of success for a given amount of effort and expenditure. Successfully implementing a project selection strategy that maximizes synergistic effects of individual projects will require greater popular support for a comprehensive plan both from within the state and at the national level. Such support will likely come about only through greater public involvement in the decision-making process of a comprehensive plan. **Louisiana's restoration goals should be better defined and more clearly communicated to the public.** This means that maps of the region and projected land-use patterns with and without various restoration projects should be circulated. Without a clarified definition of the temporal and spatial dimensions of "restoration," unrealistic expectations and disappointments are likely. The projections can be revised as additional data become available and a better understanding is developed through the adaptive management program and the science plan.

Although some inhabited areas will require relocation in order to carry out some proposed wetland restoration efforts, it will be difficult to persuade those affected by local relative sea level rise to abandon their property without a program of financial compensation and a social plan to maintain the cultural integrity of the affected communities. It is important that decisions involving relocation and compensation following Hurricanes Katrina and Rita, or in response to future events, be made in such a manner as to minimize the likelihood of additional relocation or disruption in response to future restoration efforts. The appropriate decisions and responses after major storms have to reflect a broad consensus about the future nature of coastal Louisiana and may have to include managed retreat. **Managed retreat and various restoration strategies should include early and active stakeholder participation and concurrence.** Relocation could occur either gradually with a few families at a time or at a much higher rate in areas severely affected by Katrina and Rita or future events. This is not intended to preclude reoccupation of the many areas affected by the recent hurricanes or similar events in the future. Rather, this approach is intended to minimize the potential for disrupting lives and property a second time as efforts to protect and restore Louisiana unfold in coming years.

Finally, the LCA Study calls for a long-term study of the possibility of establishing a new lobe of active delta development through a diversion near Donaldsonville, Louisiana. Termed the Third Delta, this proposed restoration feature was among a group of possible features[2] that was shown to yield limited benefits at a substantially higher cost than the projects identified for funding in the LCA Study. An alternative scenario for retention of sand and silt now lost beyond the shelf break would involve diverting the main flow of the Mississippi River toward the west of

---

[2]Groups of features, which may be made up of one or more projects, were referred to as a restoration framework in the LCA Study.

its present main channel somewhere between New Orleans and Head of Passes. An intermediate- and long-term consequence of this action would be the abandonment of the active Birdsfoot Delta by the Mississippi River. A clear benefit would be the nourishment of eroding coastal reaches to the west. Although this alternative has been widely acknowledged as possible, its feasibility, for various reasons, has not been considered seriously by USACE. Therefore, it is not yet possible to assess the potential advantages and disadvantages of Birdsfoot Delta abandonment at this time. Obviously, implementation of such a strategy would have to be accompanied by the creation of a deep navigation access channel somewhere downstream of New Orleans but upstream of Head of Passes. **Though the size of the area it would impact would still make it controversial, some consideration should be given to an alternative or companion to the planned Third Delta, such as a larger-scale diversion closer to the Gulf of Mexico, that would capture and deliver greater quantities of coarse and fine sediments for wetland and barrier island development and maintenance.**

Q: *What major questions need to be answered to support implementation of the LCA Study? Are the proposed science and technology, demonstration project, and adaptive management programs appropriately structured to fill these information gaps?*

A number of technical challenges were identified in the LCA Study that should be adequately addressed through the proposed Science and Technology (S&T) and demonstration project programs (with some modification as discussed below). However, major questions regarding the future magnitude of forces driving land loss and the acceptance by stakeholders of various large-scale projects appear unaddressed by the S&T and demonstration project programs. Both of these questions will be best answered by proceeding with the LCA Study, anticipating to the degree possible, and responding to information as it becomes available. While the knowledge gap associated with the localized role individual processes may play does not preclude successful restoration, it does emphasize the need for a robust adaptive management effort. Thus, the adaptive management program will play a major role in collecting and synthesizing data and charting new directions as appropriate. **The S&T Program requires a more explicit statement of program responsibilities and means for setting priorities; it must be integrated more effectively into the central management structure through the adaptive management process and include better representation of social sciences and ecological processes.** Additional key questions relate to the causes of land loss, recognizing that the relative role of various processes is location dependent. The future rates of loss are uncertain, and some evidence suggests that the average rate of land loss across coastal Louisiana may be decreasing. Documented rates of worldwide sea level rise and regional subsidence clearly indicate that in the absence of adequate action, land loss in coastal Louisiana will continue. If, however, the rate of land loss is indeed declining, the potential to more fully offset land loss may be greater as the rate of land building, through various proposed and future efforts, begins to approach the rate at which land is lost through erosion and subsidence.

The S&T Program envisioned in the LCA Study is an innovative and essential element that provides a mechanism for planning and assimilating monitoring results and developing adaptive management strategies. Furthermore, the S&T Program is responsible for model development and maintenance. The proposed S&T Program represents a very positive step in building needed capacity for understanding how coastal Louisiana may respond to various restoration efforts or may evolve in the absence of some of those efforts. However, it is unreasonable to expect any region to have all the necessary experience and human resources to address most effectively the problems at hand. Just as the funding of the LCA Study and its extensions includes a combination of state and federal resources, the scientific and other elements of the LCA Study should draw on the best state, national, and international talents available. Therefore, **the LCA Study should direct efforts toward capacity building that enables the program to address its stated objectives by drawing on the widest possible pool of national and international technical expertise.** Such capabilities will be especially important as strategies for restoring and protecting Louisiana, in the aftermath of Katrina and Rita, are developed and implemented.

Q: *In light of the substantial financial resources that would be required to implement the LCA Study, what are the potential benefits of Louisiana's coastal restoration to the national economy and the nation's interests? How best can these potential benefits be more fully evaluated?*

The LCA Study states that "execution of the LCA [Study] would make significant progress towards achieving and sustaining a coastal ecosystem that can support and protect the environment, economy, and culture of southern Louisiana and thus contribute to the economy and well-being of the nation." The economic analysis provided within the LCA Study and its supporting documents, however, includes only cost-benefit analyses of alternative approaches to meet ecosystem restoration objectives, as is consistent with USACE policy for evaluating projects proposed as National Environmental Restoration efforts. Evaluating the benefits of restoring coastal Louisiana in terms of national economic interests, as implied by the statement of task, would have required USACE planners to carry out analyses more consistent with proposing the effort as a National

Economic Development project. USACE officials appeared to view the efforts described within the LCA Study as falling under National Environmental Restoration as opposed to National Economic Development and, thus, did not attempt to identify and meaningfully quantify the contribution to the economy of the nation. Since the information necessary to evaluate proposed coastal Louisiana efforts in terms of the national economy is not provided in the LCA Study, there is insufficient information available for the committee to comment credibly. Carrying out such an analysis would require significant effort and resources beyond those available to the committee in the 10 months following the release of the LCA Study in November 2004. This said, some components of such an analysis can be articulated.

The LCA Study presents sufficient information about the importance of some components of the natural and built environment in coastal Louisiana (e.g., system of deep water ports, oil and gas receiving and transmission facilities, complex and extensive urban landscape, robust commercial fishery) to demonstrate that substantial economic interests are at stake in coastal Louisiana and that these interests have national significance. The immediate impacts of Katrina underscore the importance of New Orleans, and adjacent areas of the Gulf Coast, to the national economy. Establishing the true, national economic significance of efforts to restore coastal wetlands in Louisiana as proposed in the LCA Study, however, must go beyond simply identifying and characterizing these components and should include an analysis of how specific restoration efforts will preserve or enhance the value of these components (i.e., some restoration efforts may have little influence on the vulnerabilities of specific components of the natural and built environment in coastal Louisiana) and should determine how the national economy would respond to the loss or degradation of components (e.g., what is the capacity for similar components in other regions to compensate for the loss and on what time scales?). **If, as implied by the statement of task, greater emphasis is to be placed on the national economic benefits of restoring and protecting coastal Louisiana, future planning efforts should incorporate meaningful measures of the economic significance of these projects to the nation consistent with procedures normally employed to determine the value of a project or a suite of projects for National Economic Development.** As a greater understanding of the short- and long-term economic impacts of Katrina and Rita becomes available, a more meaningful effort to evaluate the national economic significance of protecting the natural and built environment in coastal Louisiana will be possible. Such information would provide an important context for decision making; however, it will still be important to understand the role wetlands play in protecting specific components of the overall system and to determine how specific restoration efforts can enhance that protection. While wetlands and adjacent barrier islands and levees are known to reduce impacts from waves, their more complex role in reducing storm surge is less well known. Surges contain multiple components, including barometric tide effects, wind stress-induced setup, wave-induced setup, and Coriolis forces. As was pointed out repeatedly in the public media during Katrina and Rita, in the northern hemisphere the eastern side of a hurricane tends to drive water northward in a counterclockwise manner. If a storm stalls off a coast for a significant period of time, it will continue to drive water onshore for a prolonged period, regardless of the nature of any intervening wetland or barrier island. Thus, the potential for reducing risk due to storm surge from a particular storm is more difficult to predict.

Conversely, the significance of the coastal Louisiana wetlands to the nation in terms of both their inherent uniqueness and the ecosystem services they provide is more thoroughly documented in the LCA Study, its predecessor reports, and the scientific literature. **Although efforts to restore and protect Louisiana's wetlands will likely provide some unknown but potentially significant protection against coastal storms and hurricanes, those efforts should not be evaluated primarily on their significance for National Economic Development.**

Q: *Does the phased approach outlined in the LCA Study provide an adequate basis to start developing a comprehensive coastal restoration plan to achieve the broad goals articulated by Coast 2050?*

The two major components of the LCA Study, a series of restoration and demonstration projects designed to be implemented over a 10-year time frame and the development of a robust intellectual infrastructure to inform future project design and implementation, are at the heart of the phased approach referred to in the statement of task. This approach has decided advantages and disadvantages. As is clear from the LCA Study, simply keeping pace with land loss in Louisiana will require an ongoing effort. Any substantial gains in the next few decades will require a robust effort, an effort that needs to be well informed by a thorough understanding of both the natural physical and ecological processes involved and the viability of various restoration techniques to address land loss at a massive scale. Establishing methods that allow projects to evolve in the face of increased understanding is prudent. Conversely, limiting project selection to those features where construction can be initiated in 5–10 years presents a significant handicap for laying the groundwork for a comprehensive, multidecadal effort.

For example, the 10-year implementation criterion resulted in the selection of projects that already existed in the USACE and the CWPPRA planning process. This time constraint precluded consideration of projects with solid potential for long-term benefits that had not yet been fully designed (precluding the initiation of construction in 5–10 years). Similarly, this criterion and the need to demonstrate solid near-term success likely precluded large-scale and innovative projects that (1) affect significant sediment delivery to the system (such as abandonment of the Birdsfoot Delta), (2) maximize synergistic effects for reducing land loss over longer time scales by the selection of strategically located or larger-scale projects, or (3) address some of the difficult issues associated with stakeholder response. While the efforts preceding the LCA Study have achieved a laudable degree of unanimity among stakeholders on the conceptual restoration plan, this unanimity will be tested by the difficult decisions associated with implementation of the larger-scale projects designed to achieve a more effective delivery of sediment, water, and nutrients over a larger area. **The project selection procedure requires more explicit accounting of the synergistic effects of various projects and improved transparency of project selection to sustain stakeholder support. Furthermore, beneficial, synergistic interaction among projects cannot be assumed but should be demonstrated through preconstruction analysis.**

It is important to note that, by definition, the activities proposed within the LCA Study are intended to lay a foundation for more effective and robust efforts to preserve and protect coastal Louisiana. By its own analysis, the LCA Study points out that constructing the five restoration features it proposes would reduce land loss by about 20 percent (from 26.7 km^2 per yr [10.3 mi^2 per yr] to 22.3 km^2 per yr [8.6 mi^2 per yr]) at an estimated total cost of roughly \$864 million (or \$39,400 per hectare [\$15,900 per acre]) over the 50-year life of the projects, not including maintenance and operational costs.

Actual land building will be experienced only in areas adjacent to the implemented projects. The significant investment represented by these projects and the efforts to develop the tools and understanding necessary to support future restoration and protection efforts will yield a substantial return of benefits only if future projects are carried out in a comprehensive manner. The funding required to carry out the activities described in the LCA Study should be recognized as the first of a funding continuum that will be required if substantial progress is to be made. **A comprehensive plan to produce a more clearly articulated future distribution of land in coastal Louisiana is needed. Such a plan should identify clearly defined milestones to be achieved through a series of synergistic projects at a variety of scales.** (While a comprehensive plan is needed, this does not necessarily imply endorsement of *the* draft LCA Comprehensive Study, which was not formally released or reviewed as part of this study.) The review detailed in this report found no instance where the proposed activities, if initiated, would preclude development and implementation of a more comprehensive approach. Conversely, many examples were identified where implementing the proposed activities would support a more comprehensive approach. **Thus, the efforts proposed in the LCA Study should be implemented, except where specific recommendations for change have been made in this report and only in conjunction with the development of a comprehensive plan.**

As the State of Louisiana and the nation begin to recover from Katrina and Rita, efforts to restore wetlands in Louisiana will likely compete with reconstruction and levee maintenance or enhancement efforts. As this report and numerous other NRC reports have pointed out, efforts to design and implement water resource projects (including environmental restoration and flood control projects) should be carried out within a watershed and coastal system context. Ongoing discussion of long-term response to Katrina and Rita underscores the need to consider restoration and reconstruction as a seamless process that should be informed by a coherent, comprehensive plan that addresses the issues raised in this report. **Therefore, efforts to rebuild the Gulf Coast and reduce coastal hazards in the area should be integral components of an effective and comprehensive strategy to restore and protect coastal Louisiana wetlands.**

1

Introduction

Let the Land rejoice, for you have bought Louisiana for a Song.
—General Horatio Gates to President Thomas Jefferson on July 18, 1803

HIGHLIGHTS

This chapter
  • Includes an overview of the history and possible causes of land loss in Louisiana
  • Discusses previous programs to address land loss and the current *Louisiana Coastal Area (LCA), Louisiana—Ecosystem Restoration Study* (LCA Study)
  • Summarizes the challenges to curbing losses and the need for understanding natural processes
  • Outlines the structure of this report

On August 29, 2005, Hurricane Katrina struck eastern Louisiana, Mississippi, and western Alabama, killing hundreds of local residents, displacing hundreds of thousands more, and causing an estimated $200 billion in economic damage. Less than four weeks later, Hurricane Rita struck easternmost Texas and western Louisiana. Although the loss of life from Rita was much less, significant destruction resulted, including the reflooding of some parts of New Orleans that were damaged during Katrina. The devastation wreaked by Katrina and Rita tragically demonstrated the risks that many coastal areas face from hurricanes and associ-

13

ated flooding. Coastal Louisiana, however, faces many unique challenges, which are the subject of this report and the proposed efforts it reviews.

The Mississippi River Delta and its associated wetlands helped to shape Louisiana's culture and economy. In addition to being a land of natural beauty and bounty, it is also home to a rich diversity of peoples. While the unique culture of New Orleans and the bayous has been the traditional magnet drawing millions of tourists to the delta, a growing appreciation for the complex wetland systems of the area is attracting increasing numbers of tourists in search of the nature and history of the area. The long, slow mingling of freshwater and saltwater that takes place between the uplands and the Gulf of Mexico has produced a rich mosaic of wetland habitats that support rare and endangered species; great flocks of waterfowl; and commercially exploited populations of furbearers, fish, shrimp, crawfish, oysters, and crabs.

Even in the aftermath of Hurricanes Katrina and Rita, Louisiana will remain a center for oil and natural gas production, transportation, and refining, and its marine fisheries are among the most valuable in the nation. The access it provides to the Mississippi River Basin also makes it a hub for shipping and navigation. The committee recognizes that this report is being released at a time when there may be many more questions than answers. Even so, the report is provided at this difficult time in the hope that its advice on restoring and protecting coastal Louisiana can be considered as part of the nation's strategy to rebuild the Gulf Coast and reduce the likelihood of future tragedies, such as Katrina and Rita.

HISTORY AND CAUSES OF LAND LOSS IN LOUISIANA

Coastal deposition along the southern edge of the North American continent has been taking place for tens of millions of years. However, the modern Mississippi River Delta as recognized today began to form when, at the end of the last ice age, the drainage of the mid-continent became integrated, creating the Mississippi River itself. Like other coastal deltas, the Mississippi River Delta plain is the product of sediment deposition and accumulation where waters of the river empty into the coastal ocean. During the late Wisconsian glaciation's peak roughly 22,000 years ago, sea level was 91–106 meters (m) (300–350 feet [ft]) lower than present. When the late Wisconsian glaciers began melting 4,000 years later, an era of sea level rise known as the Holocene transgression began, and with it came dramatic changes in the basin's hydrologic character. As the Mississippi River Basin eroded, the Mississippi River changed from a system of braided streams carrying coarse-grained sediments to a sinuous, meandering, interconnected system that carried relatively fine grain clay, silt, and sand, similar in some respects to that known today.

Earlier in its history, high rates of sediment accumulation allowed the Mississippi River Delta to build seaward, eventually forming a great accumulation of sediment that thickened dramatically to the south as it crossed the southern margin of the North American continent. The complex set of wetlands and adjacent barrier islands and levees that make up present-day coastal Louisiana represents the seaward edge of a delta plain that stretches landward onto the continent. Thus, the major source of the nutrients and sediment that sustain the natural development and maintenance of coastal Louisiana is the Mississippi River. The Holocene Mississippi River Plain is composed of six delta complexes, four of which are abandoned with transgressive barrier shorelines (Maringouin, Teche, St. Bernard, and Lafourche), and two delta complexes are active (Modern and Atchafalaya) (Penland et al., 1981; Roberts, 1998). The Chenier Plain lies west of the Mississippi River Delta where shoreline changes are linked to the complex switching process (Howe et al., 1935; Gould and McFarlan, 1959; Penland and Suter, 1989; Roberts, 1998). The Chenier Plain's mud flats prograde when the Mississippi River discharges into the western region of its plain, and the shoreline advances seaward (Wells and Kemp, 1981). When the Mississippi River discharges into the eastern region of its plain, the Chenier Plain shoreline erodes landward, forming individual chenier ridges (Hoyt, 1969). The seaward portions of the present-day delta do not rest upon the North American continent but rather upon the tremendously thick accumulated sediment along the continent's edge. As a consequence, the seaward portions of the Mississippi River Delta that make up coastal Louisiana, including New Orleans, naturally experience much higher subsidence rates than the vast majority of the North American continent. Until humans began to alter the flow of river water and the sediment it carried, the natural system was, on average, able to keep pace with natural subsidence and global sea level rise.

Over the last 10,000 years, the natural processes acting upon the Mississippi River Delta have created periodic changes in the course of the lower river that resulted in the building of new lobes adjacent to the new channel alignment, while the deprivation of sediments at the abandoned site brought about gradual land loss. As is the case for the entire delta, the lost land was composed predominantly of wetlands, though adjacent barrier islands and natural breaches may have been lost as well. During this time, some portions of the coast experienced land loss while others experienced land gain as the mosaic of wetlands, barrier islands, and natural levees shifted across the delta. However, in the last 200–300 years, human intervention has disrupted this natural process and caused widespread land loss along the entire Louisiana coastline. As people settled in the wetlands in the nineteenth and twentieth centuries, the Mississippi River was engineered to prevent switching course, and the occurrence of flows

and sediment transport through natural levies and the subsequent formation of crevasse splays[1] have been eliminated or reduced. Wetlands were
drained for agriculture, canals were dredged for navigation, forests were
harvested for building materials, and levees were constructed for flood
protection.

Oil and gas exploration, with its associated canal dredging, peaked in
the 1960s to 1980s. Canals, and the spoil banks formed during their construction, alter local water circulation patterns and sediment depositional
processes. Spoil banks impede the flow of water, causing an inundation
that may be further exacerbated by sea level rise. These processes result in
the drowning of fragile terrestrial vegetation. Freshwater vegetation is
also adversely affected by saltwater intrusion from sea level rise or along
canals that provide a route for saltwater to intrude into typically freshwater areas. Another negative effect associated with newly open water bodies includes increased wave erosion. There is also evidence that the extraction of large volumes of oil and gas has exacerbated the problems of
inundation and saltwater intrusion. Finally, the introduction of nutria (a
rodent that consumes marsh vegetation and thrives in Louisiana's wetlands) has also contributed to wetland loss in many areas.

Natural causes of land loss include subsidence due to compaction of
aging deltaic sediments, geologic faulting, sea level rise, and tropical
storms or hurricanes. Prior to human settlement, land subsidence was
naturally counterbalanced over a long period of time as the resulting low
areas tended to be flooded by the Mississippi River or tidal flows, which
in turn delivered sediments and nutrients that supported marsh growth.
Since the channelization of the river and the construction of flood protection levees, however, these materials have been discharged near the edge
of the continental shelf, far from the site where they would tend to support wetland development. Without them, the wetlands cannot keep up
with all of the natural and human-caused relative sea level rises. This
problem is compounded by the rise of global sea level, which may be
accelerating in response to global warming. Taken together, the various
natural and anthropogenic factors have resulted in a wetland loss of 62.2
square kilometers (km^2) per yr (24 square miles [mi^2] per yr) on the Louisiana coast over the 10-year period from 1990 to 2000 (Barras et al., 2003).

The cost to protect and restore coastal Louisiana is estimated at $14
billion over a 30-year period, not counting long-term operations and maintenance. Among the national and local policy issues associated with or

---

[1] Fan-shaped build-up of sediment deposited where an overloaded stream breaks through
a levee and deposits its material on the floodplain.

impacted by Louisiana's coastal land loss are fisheries, oil and gas operations, navigation, water quality, sediment management, hurricane protection, flood management, and hypoxia and other issues related to nutrient pollution. While these linkages are real, it is very difficult to estimate the financial impact of future wetland loss. The public-use value has been estimated by the State of Louisiana to be in excess of $37 billion by 2050. The loss of infrastructure has been estimated at $100 billion. Hurricane Katrina caused an estimated $200 billion in damage; the degree to which this amount would have been reduced had wetland loss in Louisiana been reduced could not be determined by this committee. Development of the infrastructure impacted by Katrina contributed to the loss of wetlands; furthermore, while the existence of wetlands is known to reduce storm wave intensity, their role in reducing storm surge is less straightforward.

HISTORY OF COASTAL PROTECTION IN LOUISIANA

The challenge of slowing the loss of coastal wetlands and adjacent barrier islands and levees in Louisiana is unprecedented. The geographic extent of these wetlands[2] and the range of natural and human forces that cause wetland degradation contribute to what would be one of the largest civil works projects in U.S. history. Extending from the Texas border on the west to the Mississippi state border on the east, a distance of 515 kilometers (km) (320 miles [mi]) and a north-south extent of 40.2–160.9 km (25–100 mi), this impacted region encompasses an area of more than 32,000 km² (12,355 mi²). In the natural undisturbed state, the Louisiana coastline was subject to regions of erosion and regions of land building as the Mississippi River switched channels, which is described further in Chapter 2. The natural, but largely impractical solution, would be to allow the Mississippi River to return to this natural cycle. As the channel switches to new regions, sediments would be distributed to the shallow coastal areas rather than the edge of the continental shelf. On the scale of centuries, this would allow new areas to stabilize and become colonized by vegetation and animals. The abandoned delta lobe would gradually be lost at a rate that would depend on relative sea level rise and storm intensity. At a

---

[2]The term "wetlands" is used throughout the report to describe the complex set of environments that are typical of the Mississippi Delta (i.e., wetlands and adjacent islands and levees). It differs from the term "marsh," in that it may include localized bodies of open water; whereas, the more restrictive term "marsh" refers to vegetated areas where the soil surface is covered by water most of the time but where vegetation extends upward above the water surface.

smaller spatial scale, crevasse splays would create changes in the distributary channels and in local regions. To recover the entire area, massive quantities of sediment and prolific generation of organic matter through growth of vegetation would be required. There have been numerous efforts to deal with the problem since the late 1960s (Box 1.1).

Left to nature, many areas of erosion and accretion would be present as the river channel switched locations on a roughly 1,000-year cycle. The river channels would be much shallower and transient in location and dimensions and thus less reliable for navigation due to the shifting shoals. Without the levees, overbank flooding would nourish the marshes but threaten human lives and property. The fundamental problem is that human interests prefer a static landscape rather than a dynamic one that supports the natural river and deltaic systems and wetlands.

Slowing wetland loss in some areas and restoring wetlands in others raises a number of basic questions that go beyond the questions of sediment and sea level—even if the significant financial resources discussed in various studies and in this report were available. For example, to what extent can a sustainable coastal configuration be achieved with the many competing interests? How are the competing interests of various stakeholders to be balanced with the overall objective of capturing the maximum amount of sediment and delivering it to areas where it will provide the greatest benefit? In addition to the engineering and ecological challenges of sediment management, slowing wetland loss involves many stakeholders with diverse and sometimes competing interests. A physical "fix" in any one area will benefit some people but could damage the interests of others either in the same locality or in a different area.

Coastal Wetlands Planning, Protection, and Restoration Act

Initial efforts to offset catastrophic land loss have been implemented under CWPPRA, which called for the development of a comprehensive Louisiana Coastal Wetlands Restoration Plan (P.L. 101-646 §303.b). In 1993, the first plan was completed and used until the 1998 release of Coast 2050. Then, in 1994, the Governor's Office of Coastal Activities Science Advisory Panel prepared a plan (Gagliano, 1994), constituted under Act 6 (LA RS 49:213 *et seq.*), for the Wetlands Conservation and Restoration Authority.

Coast 2050

Coast 2050 (Louisiana Coastal Wetlands Conservation and Restoration Task Force and the Wetlands Conservation and Restoration Authority, 1998) was developed in partnership with the public, parish govern-

Box 1.1
History of Studies and Programs Dealing with
Wetland Loss in Louisiana

1967—The U.S. Army Corps of Engineers' (USACE) Louisiana coastal area study evaluated options for mitigating wetland losses under a U.S. Senate resolution.

1978—The Louisiana State and Local Coastal Resources Management Act (Act 361, LA R.S. 49:213.1) established a state coastal management program that emphasized controlling activities that cause wetland loss.

1980—Coastal management efforts in Louisiana led to the Louisiana Coastal Resources Program, which became a federally approved coastal zone management program.

1981—Act 41 of the Louisiana Legislature special session established the Coastal Environment Protection Trust Fund and appropriated $35 million for projects to combat erosion, saltwater intrusion, subsidence, and wetland loss along Louisiana's coast.

1989—Act 6 of the Louisiana Legislature passed (in the second extraordinary session) R.S. 46: 213-214, and a subsequent constitutional amendment was approved by the citizens of Louisiana, establishing the Office of Coastal Restoration and Management and the state's Coastal Wetlands Conservation and Restoration Trust Fund (also known as the Wetlands Trust Fund) to develop and implement the Coastal Wetlands Conservation and Restoration Plan for Louisiana. Income for the trust fund is a percentage of the state's mineral revenues and varies from $13 million to $25 million annually, depending on oil and gas prices and availability.

1990—The U.S. Congress passed the Coastal Wetlands Planning, Protection, and Restoration Act (CWPPRA) (also known as the Breaux Act) to contribute federal money to coastal restoration activities. This act created a partnership between the Louisiana state government and five federal agencies (USACE, the U.S. Department of Agriculture, the U.S. Department of Commerce, the U.S. Department of the Interior, and the Environmental Protection Agency).

1993—A comprehensive wetland restoration plan (Louisiana Coastal Wetlands Conservation and Restoration Task Force, 1993) was submitted to the U.S. Congress, identifying restoration projects needed to address critical wetland loss problems in Louisiana. Section 303(b) of CWPPRA directed that this plan be further developed and be consistent with the state's existing Coastal Wetlands Conservation and Restoration Plan.

continued

Box 1.1 Continued

1994—As part of CWPPRA's request for a comprehensive Louisiana Coastal Wetlands Restoration Plan (P.L. 101-646 §303.b), the Governor's Office of Coastal Activities Science Advisory Panel released an economic plan for coastal restoration in Louisiana (Gagliano, 1994).

1995—The Barrier Island Shoreline Feasibility Study was the first large-scale feasibility study undertaken by CWPPRA to assess and quantify wetland loss problems linked to the protection provided by the barrier island formations along coastal Louisiana and to develop the most cost-effective measures to minimize future wetland loss. The study was completed in 1999 and approved by the CWPPRA Task Force in 2000.

1995—The Mississippi River Sediment, Nutrient, and Freshwater Redistribution Study was undertaken by USACE to investigate potential uses of sediment, nutrient, and freshwater resources from the Mississippi River to create, protect, and strengthen coastal wetlands. The resulting report was distributed for review in June 2000.

1997—The Louisiana Conservation Plan was approved. This plan was prepared according to Section 304 of CWPPRA and includes regulatory, nonregulatory, and educational programs the state will use to achieve no net loss of wetlands. Approval of this plan reduced Louisiana's cost share of CWPPRA projects from 85 percent to 75 percent.

1998—*Coast 2050: Toward a Sustainable Coastal Louisiana* (Coast 2050) was released as the new initiative to develop a unified effort to restore and protect Louisiana's coastal resources. This plan includes a conceptual model for achieving a sustainable coast by recommending strategies for restoration. It was developed through collaboration of federal, state, and local entities; landowners; environmentalists; wetland scientists; and the public.

1999—Investigation of the Chenier Plain is completed to develop a more holistic understanding of Chenier Plain hydrology.

1999—The 905(b) reconnaissance report (U.S. Army Corps of Engineers, 1999a) was released.

2003—*Louisiana Coastal Area, LA—Ecosystem Restoration: Comprehensive Coastwide Ecosystem Restoration Study* (draft LCA Comprehensive Study) was released in draft form by USACE and the Louisiana Department of Natural Resources as a possible comprehensive, long-term restoration plan.

2004—The LCA Study was released by USACE and the Louisiana Department of Natural Resources as a near-term alternative to the draft LCA Comprehensive Study.

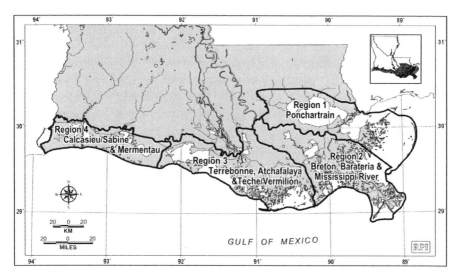

FIGURE 1.1 The regions used in Coast 2050 (Louisiana Coastal Wetlands Conservation and Restoration Task Force and the Wetlands Conservation and Restoration Authority, 1998; background map supplied by Research Planning, Inc.).

ments, and state and federal agencies under the legislative mandates described above. The plan, which divides the Louisiana coastline into four regions based on hydrologic basins (Figure 1.1), articulates a broad vision of what the citizens of Louisiana feel should be the goals of restoration and protection efforts in Louisiana. The purpose of the plan as referenced in section 303(b)(2) of CWPPRA is to "develop a comprehensive approach to restore and prevent the loss of coastal wetlands in Louisiana. Such a plan shall coordinate and integrate coastal wetlands restoration projects in a manner that will ensure the long-term conservation of the coastal wetlands of Louisiana." (Refer to Figure 1.2 to locate cities, geologic features, and projects discussed in this report.)

Coast 2050 incorporated three strategic goals for all regions: (1) to ensure vertical accumulation (of clastic sediment and organic material) to achieve sustainability, (2) to maintain an estuarine salinity gradient to achieve diversity, and (3) to maintain exchange and interface to achieve system linkages (Louisiana Coastal Wetlands Conservation and Restoration Task Force and the Wetlands Conservation and Restoration Authority, 1998). Detailed strategies to achieve these goals are included for each of the four coastal regions. In addition, during the course of the Coast 2050 planning initiative, programmatic strategies were developed to improve implementation efficiency of authorized restoration projects, effec-

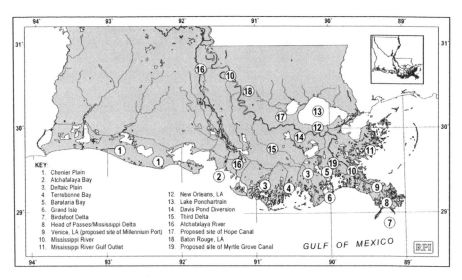

FIGURE 1.2 Map of coastal Louisiana showing major features and cities discussed in this report (background map supplied by Research Planning, Inc.).

tiveness of future restoration efforts, coordination among existing environmental resource programs, and other actions that may benefit coastal wetlands in other ways.

The draft LCA Comprehensive Study was the next step in the development of an implementation plan for Coast 2050. The goal of the draft LCA Comprehensive Study was to provide a "comprehensive program to reestablish a sustainable ecosystem along Louisiana's coast that will support and protect the environment, economy, and culture of southern Louisiana" (U.S. Army Corps of Engineers, 2003a). After reviewing the draft LCA Comprehensive Study, the U.S. Office of Management and Budget requested a more definitive and limited scope of work (i.e., one in which construction will start in 5–10 years). The State of Louisiana agreed to this narrowed scope but firmly advocated the need for a long-term plan. Currently, the State of Louisiana and USACE are pursuing implementation of a preliminary, near-term effort, which purportedly focuses on high-priority needs in the state. This near-term effort is envisioned in the LCA Study and is expected to be used to establish the intellectual infrastructure needed to carry out the more comprehensive approach. (While a comprehensive plan is needed, this does not necessarily imply endorsement of *the* draft LCA Comprehensive Study.)

LCA STUDY

Based on the U.S. Office of Management and Budget's response to the high cost and long-term elements of the draft LCA Comprehensive Study, the cooperating agencies shifted their focus to a smaller, near-term effort, which is the subject of this review. Borrowing from the plans and strategies developed in Coast 2050 and the draft LCA Comprehensive Study, the project development team narrowed the selection of near-term restoration projects to those "that can be implemented within the next 5–10 years, demonstration projects to resolve scientific and engineering uncertainty, and large-scale studies of long-range feature concepts" (U.S. Army Corps of Engineers, 2004a). The LCA Study was pulled together in a relatively short period of time as a near-term product of a much longer process that was intended to develop a comprehensive plan. As will be discussed in the following chapters, this last-minute change in focus from comprehensive strategy to near-term plan resulted in a number of disconnects between the LCA Study and the preceding efforts. Although many of the efforts proposed in the LCA Study have begun, at the date of this report, funds have not been appropriated to fully implement what is outlined in the study.

Goals and Objectives

The LCA Study encompasses the same study area and is based on essentially the same "statement of needs" as the draft LCA Comprehensive Study. However, rather than focusing on ecosystem restoration and what the end result should be, the stated objectives for the LCA Study, which focus on what needs to be done and how it will be accomplished, are as follows:

- Identify the most critical human and natural ecological needs of the coastal area.
- Present and evaluate conceptual alternatives for meeting the most critical needs.
- Identify the kinds of restoration features that could be implemented in the near term (construction starts in 5–10 years) that address the most critical needs, and propose to address these needs through features that provide the highest return in net benefits per dollar of cost.
- Establish priorities among the identified near-term restoration features.
- Describe a process by which the identified priority near-term restoration features could be developed, approved, and implemented.
- Identify the key scientific uncertainties and engineering challenges facing the effort to protect and restore the ecosystem, and propose a strategy for resolving them.

• Identify, assess, and if appropriate, recommend feasibility studies where construction starts in 5–10 years to fully explore other potentially promising large-scale, long-term restoration concepts.

• Present a strategy for addressing the long-term needs of coastal Louisiana restoration beyond the near-term focus of the [LCA Study]. (U.S. Army Corps of Engineers, 2004a)

The broad goal of Coast 2050—"to sustain a coastal ecosystem that supports and protects the environment, economy, and culture of southern Louisiana, and that contributes greatly to the economy and well-being of the nation" (Louisiana Coastal Wetlands Conservation and Restoration Task Force and the Wetlands Conservation and Restoration Authority, 1998)—has been developed into a narrower, overarching LCA Study objective "to reverse the current trend of degradation of the coastal ecosystem" (U.S. Army Corps of Engineers, 2004a). The contrast between the two goals—"sustain a coastal ecosystem that…contributes greatly to the economy and well-being of the nation" and "reverse the current trend of degradation of the coastal ecosystem"—represents a difference in viewpoint that ultimately influenced the planning process and selection of projects, because the LCA Study goal places greater influence on the implied economic benefits to the nation while the Coast 2050 goal places greater emphasis on environmental restoration.

Planning and the Plan Development Process

The agencies participating in the LCA Study were the same as those on the project delivery team[3] for the draft LCA Comprehensive Study. The plan development process is described in six phases, and the first four phases are identical to Phases I-IV of the draft LCA Comprehensive Study (see Chapter 4). Phase V for the LCA Study is to "address completeness of coastwide restoration frameworks in a Tentative Final Array," and Phase VI is to "identify highly cost-effective restoration features within the Final Array that address the most critical ecological needs" (U.S. Army Corps of Engineers, 2004a). This last phase essentially narrows the focus to select fewer projects that are more easily undertaken, use known technology, and can be started in 5–10 years. This selection process resulted in the following five near-term restoration projects:

---

[3]The New Orleans District of USACE, the Louisiana Department of Natural Resources, the Environmental Protection Agency, the Natural Resources Conservation Service, the U.S. Geological Survey, the U.S. Fish and Wildlife Service, the National Oceanic and Atmospheric Administration, academic researchers, and contractors.

- Mississippi River Gulf Outlet environmental restoration features
- Small diversion at Hope Canal
- Barataria Basin barrier shoreline restoration (Caminada Headland and Shell Island reaches)
- Small Bayou Lafourche reintroduction
- Medium diversion with dedicated dredging at Myrtle Grove

These projects, a Science and Technology Program, adaptive management, and demonstration projects are discussed in more detail in Chapter 6.

Public comments on the draft programmatic environmental impact statement and draft LCA Study were collected from July 9 to August 23, 2004. During this public comment period, six public meetings were held throughout the Louisiana coastal area, and three additional meetings were conducted in Texas, Mississippi, and Tennessee. A total of 355 people attended, and a total of 77 individuals offered oral comments at the nine public meetings. USACE received 82 comment letters postmarked within the comment period. This level of public participation is far lower than that seen for either the development of Coast 2050 or the draft LCA Comprehensive Study, in part reflecting the abbreviated time during which the LCA Study was developed.

ORIGIN AND SCOPE OF THE CURRENT STUDY

In an effort to secure additional outside technical input, the State of Louisiana and USACE initially requested that the National Academies undertake a review of the objectives envisioned and the actions proposed in Coast 2050. Once funding was in place and the committee was formed, it became apparent to both the sponsors and the National Academies that greater benefit would be derived from a review of the LCA Study. (Initially drafted as a comprehensive plan in 2003, the LCA Study reviewed by the committee was rescoped and released by USACE in November 2004.) The change in focus was approved by both the sponsors and the National Academies and is reflected in the committee's formal statement of task (Box 1.2).

The Committee's Approach

The nature of the questions comprising the statement of task places emphasis on understanding the strategic value of various components of the LCA Study. Rather than a detailed technical review of the LCA Study and its supporting documents, the committee was charged with commenting on the appropriateness of the effort in terms of broad approaches, scale of effort, and timeliness of action. In order to address the questions

Box 1.2
Statement of Task

This study will evaluate the near-term plan for the restoration of coastal Louisiana (released by the U.S. Army Corps of Engineers as the *Draft Louisiana Coastal Area [LCA], Louisiana—Ecosystem Restoration Study* in July 2004). The overall committee approach will be to examine the LCA Study and all of its components in detail. This examination will then serve as a basis for evaluating the usefulness of the LCA Study for developing and implementing a long-term comprehensive program consistent with the broad vision articulated in *Coast 2050: Toward a Sustainable Coastal Louisiana* (appended to the U.S. Army Corps of Engineers' Louisiana Coastal Area 905(b) reconnaissance report in May 1999). Specifically, the committee will address the following questions:

• Are the strategies outlined in the LCA Study based on sound scientific and engineering analyses and are they appropriate to achieve the goals articulated in the plan? What other approaches might be considered? Are adequate measures of success articulated in the LCA Study?
• What major questions need to be answered to support implementation of the LCA Study? Are the proposed Science and Technology, Demonstration Project, and Adaptive Management programs appropriately structured to fill these information gaps?
• In light of the substantial financial resources that would be required to implement the LCA Study, what are the potential benefits of Louisiana's coastal restoration to the national economy and the nation's interests? How best can these potential benefits be more fully evaluated?
• Does the phased approach outlined in the LCA Study provide an adequate basis to start developing a comprehensive coastal restoration plan to achieve the broad goals articulated by *Coast 2050: Toward a Sustainable Coastal Louisiana*?

framed by the statement of task, the committee had to understand the challenges facing existing efforts to restore and protect coastal Louisiana as well as efforts proposed in the LCA Study. This report has benefited considerably from the many informative presentations made by individuals with substantial knowledge of the history and processes associated with the Louisiana coastal area. Interaction with the National Technical Review Committee (see Chapter 4 for additional detail), including the opportunities to participate in two of its meetings and access its reports and findings, has been extremely valuable. This considerable body of work raises many overarching issues that are described and discussed in this report.

Existing commercial interests, recreational opportunities, social activities, and infrastructure needs place constraints on the remedial measures that can be carried out, thereby reducing the scope of the overall restoration that can be accomplished. For example, many existing communities are located on the relatively high ground formed by natural levees. These areas are now subsiding, becoming more vulnerable to flooding, but the prospect of future intentional sediment-laden water diversions to these areas is quite unlikely. This essentially rules out a natural approach to maintaining land areas in such locations and leaves constructed levees around the developments as the most probable future approach for flood protection. Another example is that the municipal and industrial water supply for New Orleans is obtained from the Mississippi River. With the present deep navigation channel, saltwater encroaches upstream during periods of low flow to a point where, twice in the past, consideration was given to constructing a sill to block saltwater intrusion. However, municipal and industrial water supply needs must be a factor in any proposal to divert upstream river water to carry sediments into marshes. Also, while a new Third Delta is considered a possible long-range project, it is clear that the effort to prevent the Mississippi River from following its natural, historic pattern of shifting will continue as long as such a change would jeopardize New Orleans or the viability of the port.

The size of the affected system is also an important physical constraint. In some areas close to existing waterways, sediment can be delivered via gravity by diverting sediment-laden water through spillways and siphons. Unfortunately, existing waterways do not reach much of the vast affected area, which leaves two possibilities. The first is to deliver sediment via pipeline, and the second is to accept that many of the wetlands will continue to subside with attendant wetland loss. In the latter case, as the area subsides over the years, wetlands will be replaced by more contiguous areas of open water.

A logical approach to the planning of restorative actions commences with a clear understanding of the dynamic nature of the natural system and the natural processes that create and sustain the system. Some of the natural processes are constrained by actions outside the control of efforts in Louisiana (sediment deposition in upstream reservoirs and contributions to the heavy concentrations of nutrients in river waters), and others will be unacceptable to the existing web of commercial, recreational, and societal activities and interests. Thus, there are limitations to what can be accomplished, and successful execution of the program rests as much on the recognition of these limits as on the engineering and construction of projects selected for implementation.

In summary, there are stabilized elements of the modern system of coastal Louisiana that are of critical importance to society, including the water supply for New Orleans, navigation, and a relative permanence of the present landscape and its urban and industrial infrastructure. It is worthwhile noting that the LCA Study seeks to maintain, more or less, the current landscape; however, many diverse natural landscapes have existed in the past. An example is that at one time the Mississippi River with its natural levees ended at the current Head of Passes. Were all of these past landscapes less desirable than the one that these efforts are designed to preserve?

This brief discussion illustrates the complexities facing the planners, engineers, scientists, and decision makers involved in designing and implementing the LCA Study. Broad and strong political support at the state level for the LCA Study is evident, but this can be maintained only through effective management and the achievement of early successes.

Structure of This Report

This document represents the cumulative efforts of the committee to provide answers to the questions in the statement of task. The report contains seven additional chapters (brief overviews can be found at the beginning of each chapter). Chapter 2 offers an overview of Louisiana's past and present coastal system to provide an understanding of the limitations the physical and biological systems place on restoration efforts. Chapter 3 discusses present-day conflicts and the sociopolitical limitations to efforts to return the delta to a desirable condition. Chapter 4 describes previous and ongoing efforts, such as CWPPRA, to protect and restore coastal Louisiana. Chapter 5 focuses on the planning process used to develop the LCA Study, while Chapter 6 discusses the LCA Study itself. Chapter 7 discusses knowledge gaps identified during the review of the LCA Study, which merit attention through an adaptive management process as efforts move forward. Finally, Chapter 8 provides an integrated discussion of the findings and recommended actions that should be taken by USACE and the State of Louisiana to strengthen the efforts laid out in the LCA Study to protect and restore coastal Louisiana.

2

The Historic and Existing
Louisiana Coastal Systems

HIGHLIGHTS

This chapter
- Presents the existing physical and biological constraints facing restoration and protection efforts in coastal Louisiana
- Briefly summarizes the history of the Mississippi River system over the last 20,000 years
- Reviews present constraints and modifications to the Mississippi River system

---

River deltas are coastal accumulations of sediment derived from the land and organic material that rivers have brought to the sea as well as organic matter, predominantly peat, developed on the terrestrial delta surface. The shape or morphology of a delta is determined by the delicate balance of sediment accumulation, compaction, subsidence of the seafloor upon which the sediment accumulates, and eustatic (worldwide) sea level rise. Understanding these processes, and how human activity can alter them, is an important step in understanding and anticipating the evolution of a given delta.

The prevailing shape of any given delta depends on the rates of sediment supplied by the rivers and the patterns and rates of sediment dispersal by coastal ocean processes and by gravity (Wright, 1985). The size of a delta depends, most importantly, on the annual sediment discharge of the river, but the most extensive deltas also tend to be developed where wide, gently sloping continental shelves provide a platform for prolonged organic and inorganic sediment accumulation and morphological progradation. The processes that disperse, transport, and deposit the sediment discharged by a river determine where, and at what rate, sediments accumulate or erode and thereby control the configuration of the resulting delta. This is true for both the subaqueous (underwater) component and the subaerial (terrestrial) delta, which must surmount the subaqueous deposits in order to prograde. Wright and Nittrouer (1995) argue that the fate of sediment seaward of river mouths involves at least four stages: (1) supply via river plumes, (2) initial deposition, (3) resuspension and onward transport by marine forces (e.g., waves and currents), and (4) long-term net accumulation. Different suites of processes dominate each stage. Interruption or alteration of any of these stages can impact the essential continual sediment nourishment of the delta.

The Mississippi River Delta, one of the world's most extensively studied deltas, is composed of sediments from a catchment that covers much of the continental United States (3.2 million square kilometers [km²] or 1.2 million square miles [mi²]) or 41 percent of the lower 48 land area of the United States (Environmental Protection Agency, 2005). Unlike some deltas, the Mississippi River Delta contains more subaerial material than subaqueous deposits. Mississippi River sediments generally prevail over the central portion of the wide, passive Gulf Coast continental margin (Uchupi, 1975). These sediments have accumulated on the shelf and at the base of the continental slope as the Mississippi River cone.

Past and present subsidence of the shelf and coastal plain reflects the large-scale response to the loading of these sediments. Isopach maps, which depict the thickness of sedimentary deposits based on syntheses of cores and seismic data (Coleman and Roberts, 1988a,b), show that late Quaternary deposits exceed 0.1 kilometers (km) (0.06 miles [mi]) in thickness over most of the shelf, and Holocene thicknesses are as great as 0.5 km (0.3 mi) in locations of maximum deposition.

In a recent analysis of deltaic systems of the world, Walsh et al. (in press) characterize the Mississippi River Delta as "proximal-deposition dominated," meaning that the bulk of the sediments discharged by the river were deposited close to the river mouth. In recent geologic history, a series of lobate deltaic projections were followed by avulsions, or channel switches. According to Kolb and Van Lopik (1966), at least 16 such lobes were created and abandoned in late Quaternary time (the last 20,000 years). More recent data indicate that since sea level reached its present, postglacial maximum approximately 7,000 years ago, six major lobes, including an incipient new one at the mouth of the Atchafalaya River dis-

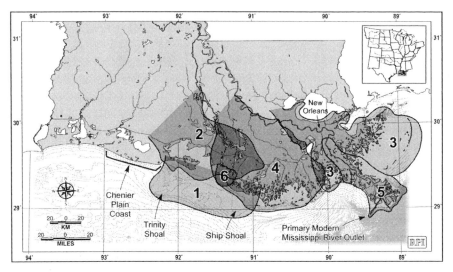

FIGURE 2.1 Past deltaic lobes of the Mississippi River (modified from Draut et al., 2005; background map supplied by Research Planning, Inc.). In order from oldest to youngest, the lobes are (1) Maringouin, (2) Teche, (3) St. Bernard, (4) Lafourche, (5) modern (Plaquemines–Balize), and (6) Atchafalaya.

tributary, have existed (Draut et al., 2005). The fifth lobe, the Balize Delta, has existed for the past 800–1,000 years, during which time it has evolved into the modern Birdsfoot Delta (Roberts, 1997).

Figure 2.1 shows the chronological distribution of the six major lobes. Abandoned deltaic lobes make up most of Louisiana's coastal plain, and over the long term, the large-scale east-to-west dispersal of sediments has been affected more by these episodic channel switching events than by oceanographic factors, such as waves and currents. Throughout the middle to late Holocene, land loss in areas recently abandoned was the norm; however, these losses were balanced approximately by accretion in areas occupied by new lobes. The alternation alongshore between areas of erosion and areas of accretion has produced a highly irregular shoreline and corresponding irregularities in the inner shelf contours. As pointed out in the programmatic environmental impact statement of the *Louisiana Coastal Area (LCA), Louisiana—Ecosystem Restoration Study* (LCA Study) (U.S. Army Corps of Engineers, 2004a), coastal erosion has been a chronic feature of coastal Louisiana throughout the late Holocene because of a combination of subsidence, delta lobe abandonment, and wave-induced erosion. However, prior to modern interventions in the form of control structures, the channel switching that allowed wetlands to be lost in abandoned areas also caused new wetlands to be created in coastal areas that were supplied by a new sediment source.

The coastal region, which is west of the deltaic plain region influenced by the delta lobe avulsions, has historically been nourished by Mississippi River sediments transported westward by coastal processes, such as waves and currents, rather than delivered directly by river mouths. This region, Region 4, is the Chenier Plain (Figure 1.1). Louisiana's Chenier Plain is a coastal morphological province characterized by a series of low-lying sandy ridges, often supporting stands of oak trees (*chêne* in French), separated by wide prograded mudflats capped by marshes (Gould and McFarlan, 1959). Episodic winnowing and reworking (by storms) of sands from within the tidal flats have formed the Chenier ridges. Mud accretion during intervening low-energy periods has produced the lower mudflats between ridges that have been subsequently colonized by salt marsh grasses. Thus, sandy, shelly eroding beaches and muddy accreting tidal flats and marshes have characterized the coast of this region through the Holocene.

The U.S. Geological Survey's report (Barras et al., 2003) on coastal Louisiana land changes describes a complex mix of land losses and gains for this region, with land loss dominating. However, Roberts et al. (2002) and Bentley et al. (2003) demonstrate that westward movement over the inner shelf of sediments supplied by the Atchafalaya River mouth, about 100–150 km (62–93 mi) to the east, has contributed to episodic progradation of the Chenier Plain. Draut et al. (2005) conclude that the eastern portion of the Chenier Plain is the ultimate sink for about 7 percent of the sediment presently discharged by the Atchafalaya River. The westward transport of this sediment apparently takes place in the form of current-driven migration of a fluid mud layer (Bentley et al., 2003). According to Bentley et al. (2003), the passage of cold fronts causes both westward and onshore movement of the mud. The resulting coastal deposition is rapid but intermittent. Therefore, there is reason to believe that sediment bypassing the current delta system, as river flow is maintained in channels leading to Head of Passes, would be distributed naturally to the west, increasing the sedimentation and accumulation rates in areas west of the current channel.

THE MODERN, ANTHROPOGENICALLY MODIFIED RIVER AND DELTA

Significant anthropogenic changes to the Mississippi River Delta began with European settlement because the natural deltaic environment was not sufficiently stable for safe or comfortable European habitation. The initial changes to the land were local and relatively small compared with modifications that commenced in the mid-nineteenth century and continue now (Syvitski et al., 2005). Changes in land use and dams upstream of Louisiana today have less of a direct impact on wetlands because levees and distributary closures have significantly separated the river from its delta.

External Changes to the Mississippi River Delta

Beginning in the nineteenth century, dams were constructed along the Mississippi River and its tributaries to improve navigation, control floods, and provide water for irrigation and electric power generation (Meade, 1995). These dams, especially in the Missouri River Basin, trap much of the suspended and bedload sediment within reservoirs. Changes in the river's hydrologic profile have led to erosion of some portions of the banks and riverbed, which remain a source of coarser material transported as bedload and delivered to the delta. Similarly, artificial levees and armoring (constructed along riverbanks for flood protection) prevent sediment introduction from bank erosion and river meandering. Countering these changes, the longitudinal profile of the river was steepened by eliminating meanders, and artificial levees kept sediment in transport within the river rather than allowing it to escape to floodplains. Published assessments suggest that the suspended sediment load of the Mississippi River and the percentage of sand in the suspended load have decreased significantly since the mid-1800s (Kesel, 1988) (Figure 2.2; Table 2.1).

For example, the suspended-sediment discharge in the Mississippi River near Baton Rouge decreased from 500 million metric tons to 200 million metric tons between 1950 and 1982 (Figure 2.2). Although organic material derived from vegetation is an important constituent of deltaic wetlands, inorganic (mineral), river-derived sediment deposition is the most important means to build new land to the level where plants can thrive. Less sediment to build land is now available than in the historic past (Kesel, 2003).

As population growth and agricultural and industrial activity have increased within its drainage basin, the nutrient load of the Mississippi River has also increased (Antweiler et al., 1995). The flux of nutrients, primarily nitrates (NO_3^-) and orthophosphates (PO_4^{-3}), increases downstream where various tributaries enter the Mississippi River. This flux is the result of numerous inputs: (1) sewage treatment, industrial wastewater treatment, and stormwater discharge; (2) automobile exhaust and fossil fuel power plant emissions; and (3) agricultural runoff from animal waste and fertilizer (National Research Council, 2000a; Mississippi River/Gulf of Mexico Watershed Nutrient Task Force, 2001). The nutrients are generally not at a high enough concentration to cause problems in the

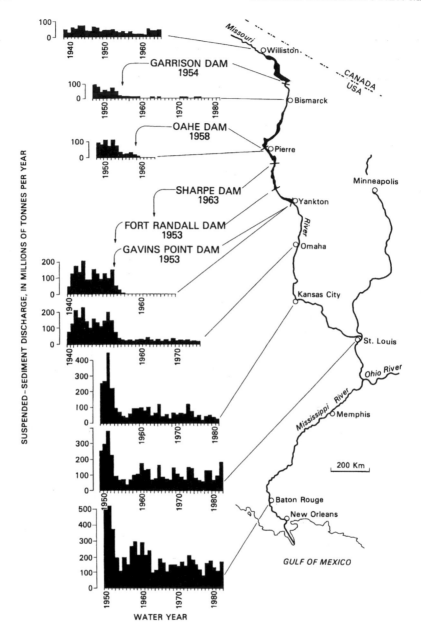

FIGURE 2.2 Annual discharge of suspended sediment at stations on the Missouri and Mississippi Rivers showing the effects of reservoirs on downstream sediment loads (1939–1982) (Meade and Parker, 1985; used with permission from the U.S. Geological Survey).

TABLE 2.1 Estimates of the Suspended
Sediment Load in the Lower Mississippi
River (at New Orleans)

Period	Suspended Sediment Transport (millions of tons per yr)
1851–1853	396
1930–1952	298
1953–1962	112
1963–1982	82

SOURCE: Kesel, 1988; with kind permission of Springer
Science and Business Media.

river. However, when this dissolved plume—in particular NO_3^-—enters
the Gulf of Mexico, it leads to phytoplankton blooms and creates a serious
problem for the nearshore ecosystem (Turner and Rabalais, 1991, 1994),
and when the dead plankton sink through the nearshore water masses,
which are stratified by vertical salinity and temperature variations, the
decay of the plankton depletes the bottom waters of oxygen and harms
pelagic and benthic communities over a large area on the continental shelf.
Evidence from cores indicates that this problem did not exist before 1900
and became acute after 1950 (Rabalais et al., 2002a).

Relative Sea Level Rise

Eustatic sea level rise occurs with respect to an absolute datum (and is
reported as a worldwide average) as opposed to relative sea level rise,
which occurs with respect to benchmarks established on land surfaces
that may themselves be sinking or rising. Relative sea level rise, which is
most relevant to the problem here, includes both true sea level rise and
decreases or increases in land elevation from subsidence or uplift, respectively. Relative sea level rise in much of the Louisiana coastal area is approximately one order of magnitude greater than the eustatic rate. A component of relative sea level rise is due to eustatic sea level rise. Recent
studies confirm that the Earth is in a period of global warming (National
Research Council, 2000b). As the world's oceans grow warmer, the ice
caps will melt, increasing the volume of the ocean. Also, even modest
warming increases the volume of water in the world's oceans through
thermal expansion (steric effects). During the twentieth century, eustatic
sea level rise occurred at a rate of 1–2 millimeters (mm) per yr (0.04–0.08
inches [in] per yr), increasing the amount of coastal land that is submerged
subject to erosional pressures and increasing the duration of flooding (U.S.

Army Corps of Engineers, 2004a). Relative sea level rise, which includes this and other components, contributes to erosional pressure and, therefore, to land loss—further complicating coastal restoration efforts. The causes and possible policy implications of eustatic sea level rise are beyond the scope of this report; however, restoration and protection efforts must address the problem, albeit indirectly through their treatment of relative sea level rise.

Internal Changes to the Mississippi River Delta

During the ninetieth and twentieth centuries, artificial levees were constructed almost to the mouth of the river and enlarged for flood protection. At the present time, about 3,620 km (2,250 mi) of levees prevent flooding of populated and agricultural areas in Louisiana and maintain navigation channels (U.S. Army Corps of Engineers, 2004a). Levees prevent overbank flooding and deny most of the delta region nutrient-rich river water and sediment. Instead, significant portions of sediment pass through the delta and accumulate on the outer continental shelf and in deeper waters of the Gulf of Mexico. The loss of this inorganic sediment from annual floods is a major underlying cause of land loss in coastal salt marshes experiencing rates of local, relative sea level rise up to 1 centimeter (cm) per yr (0.4 in per yr) (DeLaune et al., 1983, 1994; Reed, 1995).

Levees not only prevent widespread flooding; they inhibit crevasse splay formation. Crevasses occurred as large breaks through the natural levees and were a mechanism through which the river built land rapidly to the side of the main distributary channel. The average area of a crevasse splay was estimated at 1,683 km^2 (650 mi^2) (Davis, 2000), and as many as 20 were active in the period 1750–1927 (Figure 2.3). After the flood of 1927, higher levees, coupled with construction of the Bonne Carre Spillway that allowed flood waters to empty into Lake Ponchartrain, eliminated most crevasses above the Head of Passes near the river mouth. (A significant exception is the Bohemia Spillway and recently developed crevasse splays near Fort St. Phillip.) New and planned river diversions, which attempt to mimic natural processes of land building, are located largely at sites of former crevasses.

Canals and pipelines with associated spoil banks represent another major engineering impact on the delta wetlands. There are 10 major navigation canals and 14,973 km (9,300 mi) of pipelines in coastal Louisiana servicing approximately 50,000 oil and gas production facilities (U.S. Army Corps of Engineers, 2004a). The direct loss of land from these human modifications is large, but Turner (1997) suggests that most wetland loss results from secondary effects caused by canal and pipeline dredging. Canals dredged perpendicular to the coast allow saltwater to intrude

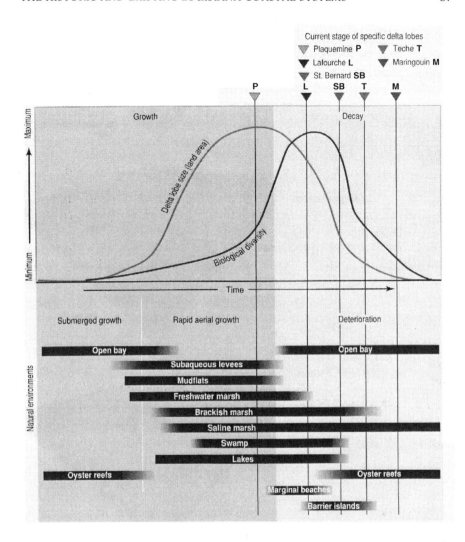

FIGURE 2.3 Graphical depiction of the growth and decay of a delta lobe or crevasse splay. Habitat and biological diversity peak in the early to middle stage of the decay phase (adapted from Gagliano and Van Beek, 1975; Neill and Deegan, 1986; used with permission from the Houston Geological Society). NOTE: A crevasse splay is essentially a small-scale subdelta lobe; even though the time scale is shorter and the space scale is smaller, the same principles apply.

into and degrade freshwater wetlands (Day et al., 2000). The Mississippi River Gulf Outlet (a 122-km [76-mi] long, man-made navigational channel connecting the Gulf of Mexico to the city of New Orleans) is often cited as a prime example of this phenomenon (Day et al., 2000; National Research Council, 2004a). Canals and spoil banks approximately parallel to the shoreline lead to ponding of water and drowning of wetlands through altered hydrology. Turner (1997) argues that most indirect wetland loss is caused by these effects. However, Day et al. (2000) claim that the influence of dredging on wetland loss varies spatially, and only 9 percent of the nondirect loss is due to altered hydrology and saltwater intrusion. These differing points of view highlight the knowledge gaps that exist with regard to the effectiveness of specific and localized wetland protection and restoration efforts.

Many towns and several large cities, including New Orleans, have been developed on the delta plain. These communities occupy large areas of former wetlands, and some are situated on the high ground of the natural levees. Where they are protected by artificial levees and by forced surface drainage (Snowden et al., 1980), no land is being lost, but urban impact on the natural landscape ranges from freshwater consumption and sewage disposal to levee construction for protection from storms and floods. Forced drainage in urban areas leads to extreme local subsidence and a need for higher levees. Industrial centers, principally those associated with oil and gas extraction and refinement, are often separate from population centers and require independent protection. Wetlands drained to allow agriculture represent lost wetlands but not necessarily lost land. Road and rail networks connecting human populations and industrial centers also destroy wetlands and require maintenance and protection from flooding. The protection of existing infrastructure from flooding was considered during the development of the LCA Study, but the impacts of Hurricane Katrina have drawn greater attention to the role wetlands play in protecting key infrastructure and urban development.

Large-scale human activities have taken many decades to have an impact, and it is difficult to separate their influence from many smaller-scale, shorter-term actions. The Old River Control Complex, for example, was completed by the U.S. Army Corps of Engineers in 1963 to prevent the Atchafalaya River from completely capturing the discharge of the Mississippi River. The structure apportions about 30 percent of the overall discharge of the Mississippi and Red Rivers to the Atchafalaya River and the remainder to the Mississippi River. Without the structure, the Atchafalaya River may have already captured most of the overall discharge. Although the Atchafalaya River would be building land faster as a result, the Birdsfoot Delta would be in a state of severe erosion, and navigation and the water supply of New Orleans area would be affected.

Much of its sediment mass would have been dispersed to the west, and a barrier island system would have begun to form from its channel, bar, and levee sands. Large hydrologic and salinity changes would also have accompanied this natural change.

Growth faults have existed in the Mississippi River Delta plain for millions of years and are a natural component of the landscape (Gagliano et al., 2003). Land sinks along growth faults on geologic time scales and can locally accelerate the rate of relative sea level rise. Morton et al. (2002, 2003a) attribute high rates of wetland loss to fault reactivation caused by pressure reductions in petroleum fields. Although the amount of land loss resulting directly from fluid withdrawal for oil and gas production is uncertain, there is a close spatial and temporal association between changes in petroleum field pressure reductions and land losses in nearby wetlands (Morton et al., 2002). As much as 43 percent of the marsh loss in southern Louisiana occurs in "hot spots" that Morton et al. (2003b) suspect are related to hydrocarbon extraction.

Shore protection structures, such as seawalls, jetties, groins, and breakwaters, are a more recent human influence on the coast and are limited to a few barrier islands and the Chenier Plain. These structures were intended to control the movement of sand and the position of the shoreline but sometimes have had unintended consequences. Areas of erosion downdrift from jetties (termed erosion shadows) are cited as a major cause of beach erosion on the Chenier Plain (Penland et al., 2004). Some prominent construction, such as seawalls near East Timbalier Island and jetties near Empire Pass Inlet, are no longer influencing shoreline sand movement because land loss has made them isolated, open water structures. The behavior of hard engineering structures during large storms in terms of their success in protecting land and property has never been proven.

The cumulative impact of human activities on sediment dispersal has generally been one of reducing sediment contributions from the Mississippi River to the delta wetlands. Instead of spreading its sediment and nutrients over a vast delta plain, the river discharges its reduced sediment load directly onto the outer continental shelf and upper slope. Land building (outbuilding) and delta maintenance (upbuilding) (Coleman et al., 1998), the large sedimentary processes through which the Mississippi River Delta was built, no longer occur. Instead of following the past history of delta growth, avulsion, and destruction, the modern Birdsfoot Delta has extended across the continental shelf.

The coastal physical oceanographic regime of the Louisiana shelf is presently scaled and constrained by the elongated Birdsfoot Delta, which creates a barrier to east-west currents (Wiseman and Dinnel, 1988). By blocking alongshore flow, this active delta lobe causes shelf circulation to be divided into two cells: one to the east and one to the west of the

Birdsfoot feature. The most recent and comprehensive analysis of the shelf circulation in this region is that of Smith and Jacobs (2005) who use extensive sets of current observational data with the governing dynamic shallow water equations. Results show that depth-averaged flows to the west of the Birdsfoot Delta during spring and summer are weak and variable but generally set from west to east. Stronger and more persistent flows from east to west prevail during autumn and winter. These flows are essential to the delivery of suspended sediments to the disappearing wetlands of Regions 2 and 3 (Figure 1.1). Research indicates that east-to-west migration of a layer of fluid mud on the Louisiana inner shelf plays an important role in nourishing the coast of southwestern Louisiana and in moderating the impact of storm waves on the coast (Huh et al., 2001; Bentley et al., 2003; Sheremet and Stone, 2003; Stone et al., 2003). Human intervention, particularly by maintenance activities or potential changes in management strategy of the Birdsfoot Delta system, has a significant influence on the coastal circulation and transport processes in this region.

Future Scenarios: Desirable Versus Attainable

The ideal future condition of the Mississippi River Delta would be one that achieves the goal of *Coast 2050: Toward a Sustainable Coastal Louisiana* (Coast 2050)—a sustainable "coastal ecosystem that supports and protects the environment, economy, and culture of southern Louisiana" (Louisiana Coastal Wetlands Conservation and Restoration Task Force and the Wetlands Conservation and Restoration Authority, 1998). To be "sustainable" requires that Mississippi River water, nutrients, and sediment be effectively spread across the wetlands of the delta. Modern human occupation requires that the sediment flow occur without river flooding or meandering in inhabited areas. Restoration to achieve this goal will require a new and significant system to convey water and sediments from the river water around inhabited areas to where it is needed. In the absence of natural avulsion, this conveyance system will require numerous large diversion structures, or a controlled—but more natural—evolution must be allowed to occur; both approaches require that pathways for water flow be created or expanded.

River sediment is essential to allow wetlands to keep pace with subsidence and sea level rise, and freshwater is needed to maintain salinity gradients. To a degree, these objectives are achieved through diversions, but canals and channels that allow saltwater to penetrate into freshwater and brackish water wetlands must be blocked. Since many of these channels are related to essential navigation needs, in an ideal world a new waterway system that separates fluvial processes from navigation would be required to protect salinity gradients.

Under conditions ideal for wetland maintenance, Mississippi River waters will disperse through wetlands before entering the ocean. Wetlands contribute to the removal of nutrients by plant uptake of soluble material and flow reduction, which enhances the deposition of organic and inorganic sediments. In surface layers, nitrification occurs and is followed by denitrification as the nitrates are carried to anoxic zones in sediments or in suspended sediment regions near the bottom. Phosphates are bound to sediments, and both phosphates and nitrates are taken up by plants. Wetlands also promote flow reductions, which enhance nutrient removal and the deposition of organic and inorganic sediments.

In the normal course of delta evolution, barrier islands move and eventually drown, and the marshes landward of them erode, supplementing the sediment supply for islands down-current through the alongshore drift. This is a part of the natural system that is not considered desirable from the perspective of wetlands restoration if the rate of wetland loss exceeds the formation of new wetlands. Thus, some maintenance of the existing barriers, or other structures, is necessary to protect some areas and to slow the loss of wetlands in other areas. Ideally, these would be self-sustaining or require little human maintenance (Reading, 1978), a characteristic that may be difficult to achieve.

With a larger and growing delta to buffer populated areas from storm waves and surges, public safety would be enhanced. Although this ideal state is a goal of the LCA Study and follows on the broad suggestions of Coast 2050, "restoration," in the strictest sense, cannot be achieved by what is proposed in the LCA Study or any plausible group of projects today. The Mississippi River Delta is inherently dynamic and large. Even maintenance of the status quo would require unreasonable quantities of sediment to travel great distances at unreasonable cost. No reasonably scoped effort will bring back the Mississippi River Delta of historic times. Those responsible for restoration efforts in Louisiana have to clarify that if these projects are executed successfully, the future delta will contain all of the landform types that exist today; however, those landforms will be smaller in size, and some will be located in different places than today. To conserve resources and focus effort where it will be most beneficial, some presently inhabited regions may have to be abandoned or relocated. If this is undertaken in a carefully planned manner that views processes on the scale of decades rather than years, the impacts to individuals and communities can be minimized.

Although the future map of Louisiana likely will not look like earlier versions, many currently nonfunctional deltaic processes can be restored, and people can live safely in a more sustainable environment than exists today. In short, Louisianans can draw their own future map of the state that enhances natural processes, ensures sustainability over decadal time scales, and protects the key values and infrastructure. The Chenier Plain, for example, can never be restored to a previous state and sustain a large human population. Growth of the Atchafalaya River Delta will eventually bring mud to the west, changing the essential character of the coastal areas in the area (i.e., sandy beaches replaced by finer-grained, muddier systems). Some areas in Regions 2 and 3 (Figure 1.1) that are too far from sediment sources to be rebuilt may become the bays and open water of tomorrow so that new land can be built to protect preferred ecosystems and population centers elsewhere. Furthermore, erosion and coastal inundation by marine water represents an important phase of delta development when submerged coastal habitats may be at their peak in productivity. Most of the seaward areas of Region 1 (Figure 1.1) will apparently be abandoned since no land-building efforts are directed to the Chandeleur Islands or the wetlands landward of them. The selection of projects now, and in the future, should be viewed as an effort to balance specific mechanisms that will draw the new map of Louisiana and not efforts that will bring back the past.

THE FUTURE LOUISIANA COASTAL SYSTEM

The Mississippi River Delta has changed constantly throughout history but experienced net growth in land area until human activities inhibited delta-building processes. The extent of human activities and population growth in Louisiana has exacerbated land loss associated with natural processes and placed the current population and its infrastructure at risk from storms and land erosion. The natural and anthropogenic processes contributing to net land loss in coastal Louisiana are significant and pervasive and have been operating for decades. Furthermore, achieving no net loss may be problematic because of the limited sediment supply; the large affected area; and the extensive social, political, and economic impediments. It is not possible to restore the earlier extent of the delta or to maintain the present status of coastal Louisiana. However, through the selection of appropriate projects and with abandonment of areas that economically cannot be saved, a sustainable delta environment and management strategy can be achieved. Additionally, the LCA Study needs to convey the message that the map of Louisiana will change in the future. **To achieve this, the development of an explicit map of the expected future landscape of coastal Louisiana should be a priority as the implementation of the LCA Study moves ahead.** Such an explicit declaration of the proposed "end state" of restoration efforts in Louisiana provides an important performance metric. Development of such a map will also require meaningful stakeholder involvement and the commitment of decision makers at all levels of local, state, and federal government.

3

Conflicts and Limitations to Achieving Goals

HIGHLIGHTS

This chapter
- Reviews the social and political constraints that the efforts outlined in the *Louisiana Coastal Area (LCA), Louisiana—Ecosystem Restoration Study* (LCA Study) will likely encounter
- Examines the role of area size and associated sediment delivery costs to counter subsidence
- Discusses activities within the Mississippi River watershed that contribute to land loss in Louisiana

---

Saving Louisiana's coastal region is a very complex and—in planning parlance—a "wicked" problem (see Chapter 5). The actions that must be taken to restore the coastal area will have to be bold, massive, costly, and continuing. The projects will challenge the technology capabilities of the U.S. Army Corps of Engineers (USACE), the Louisiana Department of Natural Resources, and the local parish governments to balance competing interests. The inherent nature of the solutions being proposed will come into conflict with the ways things are being done now. This chapter explores these challenges.

LAND LOSS PATTERNS AND PROPOSED
SEDIMENT DISTRIBUTION

Land loss is ubiquitous, occurring both in interior areas and on the edges of water bodies, including the Gulf of Mexico. Any solution approaching a large-scale or optimal restoration will encounter conflicts with navigation, flood control, oil and gas, and other land uses on the one hand and the need for large-scale redistribution of Mississippi River freshwater and sediment on the other. One of the most dramatic and long-term examples of this conflict is the dam placed across the Bayou Lafourche distributary in 1904 as a flood protection measure for Donaldsonville, Louisiana. While fulfilling its authorized purpose to prevent flooding in the city, it also reduced the natural flows of freshwater and sediments to the Barataria and Terrebonne Basins. Prior to dam construction, flows amounting to approximately 15 percent of the Mississippi River flows (Kesel, 2003) had nourished the wetlands and maintained elevations relative to sea level rise (U.S. Army Corps of Engineers, 2004a). According to the report of the Governor's Office of Coastal Activities Science Advisory Panel Workshop (Gagliano, 1994), Terrebonne and Barataria Basins each lost almost 30 square kilometers (km^2) per yr (11.6 square miles [mi^2] per yr) between 1978 and 1990.

Thus, flood control contributes to land loss, and reversing this land loss will require reflooding the area in order to preserve human habitation and agricultural productivity. Because of the extent of observed land loss across the entire Louisiana coastal area, it is clear that the constraints of existing development and the need for a minimum amount of water in the Mississippi River will limit the amount and location of any restoration. An accepted constraint of the LCA Study is that the Mississippi River switching, as would occur under natural conditions, will not be allowed. Also, the minimum Mississippi River flows must be sufficiently large so that the industrial and municipal water supply for New Orleans can be maintained (U.S. Army Corps of Engineers, 2004a).

An average subsidence of 0.25 centimeters (cm) per yr (0.1 inches [in] per yr) results in an annual volumetric deficit of 75 million cubic meters (m^3) (98 million cubic yards [yd^3]). For stability to be maintained, this volume must be replaced by a combination of siliciclastic and organic matter. Considering two ratios of siliciclastic to total matter (1:10 and 1:4), the two associated annual volumes of mineral sediment required are 75 million m^3 (98 million yd^3) and 18.8 million m^3 (24.6 million yd^3). Annual delivery costs, if distributed over the entire coastal area, would be $450 million and $1.13 billion, respectively, based on an average slurry pump distance of 120 kilometers (km) (74.6 miles [mi]) and a cost of $0.50 per m^3/km ($1.05 per yd^3/mi). The volumes of siliciclastic sediment range from 6 to 14 percent of the total combined suspended sediment delivered by the two rivers and, thus, appear feasible. However, the delivery costs indicate that adequate sediment delivery needed to balance subsidence can likely be accomplished by artificial means (pipeline transport) but only to a portion of the subsiding area. It is emphasized that these costs do not include rights of way, delays, or costs due to legal challenges. This underscores the need for the LCA Study to identify, at an early stage, those projects that can be undertaken successfully; to convey this information to the general public; and to establish reasonable expectations for accomplishments. The LCA Study also needs to focus on large-scale delivery of sediments by natural means, such as the Third Delta, to help capture economies of scale and minimize cost. (See Chapter 6 for additional discussion of the Third Delta.)

Land Building Versus Restoration

The concept of restoring a wetland to some vegetative status and function of an earlier time is contrary to recognizing the coast as a dynamic system, and the landscape is always changing under natural conditions. Restoration could also be used in the dynamic or rate-of-change sense. In this case, a wetland could be deemed "restored" if the rate of land loss (or gain) was put back to that of some earlier time. Obviously, it would take many more small diversions, such as those typically carried out under the Coastal Wetlands Planning, Protection, and Restoration Act (CWPPRA), to "restore" southern Louisiana if that means replacing the more than 4,920 km^2 (1,900 mi^2) that have been lost over the past century rather than reducing or eliminating the current rate of land loss. Since wetland functions would be restored and not the wetland itself, using the word "restoration" throughout the LCA Study invites misinterpretation and can create public expectations that are not attainable. If restoration implies recreation of land, where and when become key questions.

Inhabited Areas Versus Need to Increase
Elevation to Counter Subsidence

The entire coastal Louisiana area is subsiding, although at various rates in different regions. Without substantial engineering, continued subsidence and sea level rise will make much of the southern delta uninhabitable at some future time. An alternative, short-term solution would be to build ring levees; another would be to raise all local buildings. The only long-term solution seems to be to abandon these areas and find alternate sites for residents. These solutions are up against geologic time scales in the building, abandonment, and disappearance of deltas. Humans try to fix things on human time scales. There appears to be a need to adjust problem solvers' thinking to the reality of longer-term natural processes.

STAKEHOLDERS WITH CONFLICTING INTERESTS

There have been large-scale efforts to promote public participation during the development of *Coast 2050: Toward a Sustainable Coastal Louisiana* (Coast 2050) (65 meetings with 1,756 total attendees). All 20 parishes in the Louisiana coastal area passed resolutions of support for Coast 2050. With this in mind, the LCA Study's planning documents and programmatic environmental impact statement have noted the concerns of different interest groups.

Navigation Versus Restoration:
The Example of the Mississippi River Gulf Outlet

The Mississippi River is an important waterway for shipping goods into and out of the U.S. heartland. Shipping lanes of coastal Louisiana are a vital link between producers and consumers throughout the world. A complex of deep draft ports includes the Port of South Louisiana, which handles more tonnage than any other port in the nation, and the most active segment of the nation's Gulf Intracoastal Waterway. The principal commodities moving on the Gulf Intracoastal Waterway include chemicals, petroleum products, and crude oil. The Mississippi River Gulf Outlet (MRGO) was excavated in 1963 to shorten the trip from the Gulf of Mexico to New Orleans docks by five hours and to bring economic development to eastern New Orleans and St. Bernard Parish docks. However, in 2003, MRGO averaged 5.2 ship passages per day (U.S. Army Corps of Engineers, 2003b). USACE let maintenance dredging contracts in 2003 for more than $22.8 million, representing a cost of $12,000 per ship passage that year (U.S. Army Corps of Engineers, 2003c).

Also, MRGO has resulted in tremendous environmental damage, including saltwater intrusion, land loss, and worsening the effects of wave damage during hurricanes and storms. In the past 40 years, erosive forces have caused MRGO to increase in width from the initial 182.9 meters (m) (600 feet [ft]) to 609.6 m (2,000 ft) (an average of 10.7 m per yr [35.1 ft per yr]), causing a land loss of more than 81 km^2 (31.3 mi^2). MRGO may also pose an increased storm surge threat (Tardo, 2003; Lake Pontchartrain Basin Foundation, 2005).[1]

---

[1]At the time of this writing, some speculation about the role MRGO may have played in enhancing the storm surge during Hurricane Katrina and subsequent flooding of St. Bernard Parish had been put forward. What is known is that the levee that separates MRGO from St. Bernard Parish was topped during Katrina, as predicted in advance by some storm surge models.

In 2000, a plan to modify MRGO by halting the channel maintenance dredging and thereby eliminating deep draft vessels was proposed by the environmental subcommittee of the MRGO Policy Committee. The channel would be open to recreational boaters and commercial fishing vessels and would have a projected depth of 3.7 m (12 ft), which is considerably less than the present 11 m (36 ft) (both below mean low Gulf of Mexico datum). In addition, water control structures, including floodgates, locks, weirs, and sills, would be strategically built along MRGO to reduce the potential for saltwater influx into the marshes and bayous from the current MRGO channel, thus reducing the potential for storm surges and saltwater intrusion. This plan was endorsed by the public at the Coast 2050 meetings and has been a part of the Lake Pontchartrain Basin Foundation's comprehensive plan since 1991 (Lake Pontchartrain Basin Foundation, 1995).

The navigation industry might support "closing" MRGO if it could be confident that shipping and barge traffic displaced by the modifications would be accommodated elsewhere (Caffey and Leblanc, 2002). From the industry's perspective, this means, at a minimum, that the Port of New Orleans would complete the expanded container facilities on the Mississippi River and improve ship access through the Inner Harbor Navigational Canal. The ideal situation for the shipping industry would be construction of the "Millennium Port" (Caffey and Leblanc, 2002). Some groups advocate closing MRGO without further studies.

Oil and Gas

Oil and gas production is very important to the economy of Louisiana. With more than a million production wells drilled in Louisiana and offshore, Louisiana ranks first in the production of crude oil—second only to Texas in the production of natural gas—and supplies roughly one-quarter of the natural gas used in the United States. The direct and indirect impact of the oil and gas industry on the Louisiana economy totals $92.6 billion; it supports more than 341,000 direct and indirect jobs and is responsible for more than $12.2 billion in household earnings (13 percent of total earnings in Louisiana) (Scott, 2002). The oil and gas industry provides nearly one-quarter of total revenues collected by the state. As oil and gas resources within the state have been depleted, the industry has slowly moved offshore, discovering and developing even more significant resources in deep water environments of the prodelta (e.g., the deepest portions of deltaic sedimentary deposits, the prodelta marks the oceanward transition to normal environments of the Gulf of Mexico). In 2003, Louisiana received roughly $29.6 million in revenues (royalties, rents, and bonuses) from the federal government as its share of offshore leases for oil and gas production. In 2001, Louisiana received $40.6 mil-

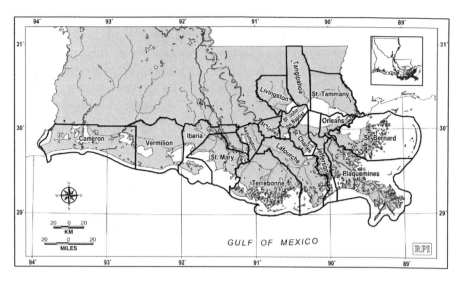

FIGURE 3.1 The 17 parishes that were part of the socioeconomic analysis in the Coast 2050 programmatic environmental impact statement (background map supplied by Research Planning, Inc.).

lion in revenues from federal offshore mineral leases and $1.4 million in federal onshore leases (Minerals Management Service, 2005). With more than 62,000 onshore wells in 17 parishes (Figure 3.1), Louisiana received roughly $419 million in royalties in 2003 (Table 3.1).

In the development of oil and gas wells, thousands of miles of canals have been cut through the marsh to aid exploration, to sink wells, to install pipelines, and to service the oil and gas industry. In order to restore the marsh areas, most of the canals have to be closed (or plugged) so that saltwater can be prevented from entering the freshwater marshes, creating a continuous marshland area. The filling will be either through diversions from the Mississippi River or by using dredge material from shipping waterways or from offshore. The end result will be reduced access to the oil and gas production wells, a problem that becomes less of an issue as the production from these wells decreases. As discussed earlier, subsidence of the Louisiana coastal area may also be affected by the extraction of oil and gas found below the marshes.

Commercial and Recreational Fishing

Commercial and recreational fishing are two of the most significant industries in the economy of the study area. Excluding Alaska, Louisiana

TABLE 3.1 Louisiana Oil and Gas Profile by Parish in 2003

Parish	Number of Wells	Royalties (dollars)[a]	Gas Severance (dollars)[b]	Oil Severance (dollars)[b]
Assumption	492	2,765,265	1,661,481	2,051,860
Cameron	5,555	23,426,231	9,751,601	11,555,453
Iberia	2,178	11,109,959	1,582,529	17,405,109
Jefferson	1,746	16,872,136	860,236	8,743,873
Lafourche	6,884	25,460,199	5,193,455	30,155,887
Livingston	227	86,598	135,764	559,971
Orleans	8	0	0	822
Plaquemines	25,373	152,323,581	16,486,586	72,066,747
St. Bernard	1,777	56,306,199	10,201,984	3,785,010
St. Charles	901	259,822	419,808	3,100,211
St. James	288	63,599	339,467	356,029
St. John the Baptist	65	0	41,119	166,538
St. Mary	5,533	54,746,858	8,008,321	12,539,741
St. Tammany	4	0	12,345	0
Tangipahoa	18	0	-78,748	1,723
Terrebonne	6,459	55,402,555	15,888,415	59,095,001
Vermilion	4,991	173,900	429,895	2,644,669
Total for these parishes	62,499	398,996,902	70,934,258	224,228,644
Louisiana totals	447,402	419,218,358	152,887,515	323,593,871
Percent of state totals	14	95	46	69

[a]Royalties are fees paid to the state from oil and natural gas production taking place on state-owned lands and bottoms of waterways. This amount is negotiated with each lease.
[b]Severance tax is collected on oil and gas production within the state and out to the three-mile offshore boundary. It is 12.5 percent of the value of the product.

SOURCE: Data complied from Louisiana Mid-Continent Oil and Gas Association, 2003.

produced the nation's highest marine commercial fish landings (about $343 million), minus mollusk harvest, such as clams, oysters, and scallops (National Marine Fisheries Service, 2003). Louisiana produced 37 percent of the nation's oyster needs and 26 percent of the blue crabs in 2001. Shrimp landings in Louisiana were approximately 56,700 metric tons (125 million pounds) during 2001, representing more than 45 percent of the total U.S. landings. Commercial fishing supports approximately 31,400 jobs in Louisiana (U.S. Army Corps of Engineers, 2004a). In 2001, recreational fishing expenditures in Louisiana were $703 million (U.S. Department of the Interior and U.S. Department of Commerce, 2003). The large expanse of coastal wetlands and estuaries provides support during the critical life stages of important commercial and recreational species. With the change in habitat comes a change in finfish and shellfish species.

As restoration proceeds, the goal is for open water to become saline marsh and then freshwater marsh; the change in habitat types will result in a change in abundance of finfish and shellfish species or in the geographic range of a particular species. Commercial and recreational fishermen are concerned that the change in the salinity regime associated with a freshwater diversion would cause loss or displacement of current commercially and recreationally valuable fishery species. The diversions may also increase the amount of nutrients supplied to lakes and bays. Increased nutrients create the possibility of algal blooms that contribute to the formation of hypoxic zones.

Mississippi River water and sediments are proposed for use in rebuilding marshes and land. There may be concerns about the quality of these resources (e.g., are they appropriate for use in restoration, are they contaminated) and concerns that the restoration processes may intensify the problems within the receiving areas or adversely affect human health through consumption of contaminated finfish and shellfish.

Recreation and Tourism

The abundance of natural and cultural resources in the Louisiana coastal region supports a diversity of recreational activities. The coastal area serves as a home to thousands of wildlife species that attract individuals for many types of tourism. Along the coast are opportunities for fishing, hunting, boating, swimming, camping, bird watching, crabbing, and crawfishing.

The rich historical and cultural traditions of southern Louisiana give rise to hundreds of local and regional festivals focused on the harvest of rice, sugarcane, shrimp, crawfish, oyster, and alligator. Other festivals celebrate the birds that pass through the state, as well as the Cajun, Native American, African American, and European cultures and heritages of the people. These festivals are economically significant to coastal communities, attracting many noncoastal visitors.

In 2000, 802,000 Louisiana residents and 38,000 nonresidents participated in wildlife watching and spent $168.4 million, and in 2001, hunting expenditures in Louisiana were $466 million (U.S. Department of the Interior and U.S. Department of Commerce, 2003). Overall, people traveling to Louisiana spent approximately $8.1 billion in 2001, supporting more than 113,000 jobs in the state with an annual income of about $1.8 billion. Tax revenues associated with recreation and tourism in Louisiana were about $1.1 billion for all levels of government (U.S. Army Corps of Engineers, 2004a). With major river diversions flooding areas, other wetland areas closed to entry for replanting, and canals being plugged, there will be an impact on this component of the economy. Tourism activities may change spatially in response to these restrictions, but losses in some areas may be offset by gains in others, especially in the short run, while all areas may gain in the long run.

Agriculture and Rural Economy

Agriculture in the 20 coastal Louisiana parishes contributes $410.8 million in market value of agricultural crops, or 22.6 percent of Louisiana's total crop value of $1.8 billion (U.S. Department of Agriculture, 2002). The rich deltaic soil and mild climate are conducive to producing a wide variety of crops. Crops are irrigated with water from local bayous, and with increasing water salinity, crop production has declined.

In the 1850s, there were several hundred farms and sugar plantations along Bayou Lafourche and its neighboring waterways, and the sugar crop, with the improvement of refining techniques, dominated both the economy and the culture of southeastern Louisiana. Most of the large sugar plantations, however, are now located farther up Bayou Lafourche on land that is better drained than the low, marshy, coastal area (U.S. Army Corps of Engineers, 2003a). The number of farms in the coastal area is decreasing, and in general, the size of the farms is increasing, which suggests that the family farm is disappearing (Table 3.2). The number of farms in the coastal area is decreasing (except in Plaquemines Parish, which is increasing). The population of St. Tammany Parish has increased 48 percent between 1999 and 2004, and Livingston has grown 50 percent in the same period. Both parishes show a decrease in the number of farms; Livingston Parish farms have decreased by 21 percent in size and decreased by 37 percent in average production (Table 3.2). The size of farms in St. Tammany has increased by 10 percent between 1997 and 2002, and the value of the production during that period has decreased by 3 percent. As agriculture declines in importance due to the decline in agricultural production and in the numbers of people employed, agriculture, as an economic sector, will lose political strength, and the focus of land restoration may change from protecting farms to protecting communities.

Today the major crops are still sugarcane (37 percent of the nation's sugar), rice (20 percent of the nation's rice), and soybeans. (Where the crop values are high, the largest commodity is sugarcane, and where crop values are low, but subsidies are high, the commodity is generally rice.) Much of this agricultural land is considered prime farmland and protected under the U.S. Farmland Protection Policy Act of 1981 (P.L. 97-98). Optimal wetland restoration benefits may require major river diversions to flood prime agricultural land, and roads needed to transport crops to market could also be impacted.

Since the area around New Orleans is urbanizing, the previously rural areas are decreasing. The proposed projects will have the most impact

TABLE 3.2 Changes in Farming in Louisiana's Coastal Area

Parish	Number of Farms in 2002	Percentage Change in Farm Number Since 1997	Average Farm Size in 2002 in km² (mi² in parentheses)	Percentage Change in Average Farm Size Since 1997	Average Production in 2002 (dollars)	Percentage Change in Average Production Since 1997
Assumption	105	−20	2.5 (1.0)	22	343,231	30
Cameron	409	−16	2.5 (1.0)	16	15,352	−33
Iberia	340	−13	1.3 (0.5)	16	155,064	15
Jefferson	52	−36	30.9 (11.9)	27	32,345	−38
Lafourche	405	−16	1.5 (0.6)	30	66,067	−6
Livingston	451	−6	0.3 (0.1)	−21	155,684	−37
Orleans	NA	NA	NA	NA	NA	NA
Plaquemines	192	19	0.7 (0.3)	−27	34,391	4
St. Bernard	24	−31	NA	NA	15,654	50
St. Charles	62	−26	0.6 (0.2)	−44	85,533	37
St. James	69	−14	3.1 (1.2)	31	386,211	14
St. John the Baptist	34	−11	2.6 (1.0)	147	170,345	81
St. Mary	99	−18	3.0 (1.2)	4	358,871	7
St. Tammany	603	−4	0.3 (0.1)	10	20,748	−3
Tangipahoa	1,065	−10	0.4 (0.2)	0	53,275	0
Terrebonne	156	−8	1.4 (0.5)	4	98,230	11
Vermilion	1,116	−10	1.3 (0.5)	17	47,046	−20

NOTE: If no value is given, it was not available (NA).

SOURCE: U.S. Department of Agriculture, 2002.

on Jefferson, Lafourche, Livingston, Plaquemines, and St. Bernard Parishes. Livingston's largest agricultural crop is broilers and other meat-type chickens; Plaquemines biggest crop is oranges, and it has 7,000 head of cattle. St. Bernard has most of its farmland in forage, and Lafourche has 107 km² (41 mi²) in sugarcane.

Although there is a center of urbanization, the rest of the Louisiana coastal area is rural. Table 3.3 shows how employment is spread among industries in 17 parishes. The diversification of the southern Louisiana economy increased after the local recession of the late 1980s as resources were channeled from the oil and gas industry into other areas, including tourism. However, the table is somewhat deceptive in that a strong dependence on the oil and gas industry still remains.

What is striking about the housing conditions is the large number of mobile homes in Louisiana's coastal area—a high of 31.5 percent in Plaquemines Parish and a low of 0.3 percent in Orleans Parish. Table 3.4 also illustrates the rural counties' low number of housing units and high percentage of mobile homes. The suburban counties have a high percentage of owner-occupied housing. There is also a large discrepancy between total housing units and the number of occupied housing units. Orleans, as one might expect, has a higher percentage of rental units than owner-occupied units.

Flood Control and Land Building

Nearly two million Louisiana residents live in the coastal zone, and the culture and socioeconomic structure of the population has evolved to depend on the presence and productivity of wetlands, as well as the flood protection provided by levees. Community and regional growth would not have been possible without construction of an extensive network of levees and floodgates along the Mississippi River for flood protection. For instance, between 1735 and 1927, New Orleans flooded nine times due to levee breaks or crevasses upstream (U.S. Army Corps of Engineers, 2003a). Numerous lesser flood control and hurricane protection projects have also allowed development. Except for the recent hurricanes, flood losses occurred mainly as a result of rainfall events. These damages increased as development continued, thereby reducing flood storage area while the assets at risk of flood damage became greater. Between 1978 and 2001, a total of $1.08 billion was paid by the Federal Emergency Management Agency for Louisiana flood claims. The majority of damage has taken place in the most densely developed parishes of the coastal area: Orleans and Jefferson.

Diverting river flow to build land and marshes requires breaching levees and flooding areas. The large amount of money—up to 35 percent

TABLE 3.3 Employment Numbers in Coastal Louisiana by Industry in 2000

Parish	Agriculture, Forestry, Fishing, Hunting, and Mining	Construction	Manufacturing	Wholesale Trade
Assumption	712	1,411	1,354	173
Cameron	696	470	295	143
Iberia	3,896	2,094	3,042	1,228
Jefferson	4,059	16,353	17,663	9,910
Lafourche	3,066	2,970	4,928	1,295
Livingston	478	6,993	4,880	1,626
Orleans	1,996	9,478	9,925	4,885
Plaquemines	1,211	715	899	368
St. Bernard	563	2,700	3,165	1,350
St. Charles	355	2,136	3,876	999
St. James	201	611	2,012	197
St. John the Baptist	209	1,833	2,719	666
St. Mary	1,781	1,751	2,468	602
St. Tammany	2,265	8,044	6,866	3,551
Tangipahoa	1,516	3,638	4,436	1,568
Terrebonne	4,916	3,248	3,437	1,668
Vermilion	3,435	1,660	1,410	773
Total	31,355	66,105	73,375	31,002
State totals	78,167	145,850	187,499	65,247
Percentage of state totals	40.1	45.3	39.1	47.5

of a project's total cost—for land purchase and relocation of homes, residences, and businesses reflects the constraints of flooding currently inhabited or agriculturally productive lands (U.S. Army Corps of Engineers, 2004a). Gaining support from those property owners and business operators who will be adversely impacted by restoration efforts will continue to be a challenge. Continuing discussion with these stakeholders will be even more important as relocation and reconstruction begin in the aftermath of the hurricanes.

Urbanization

The urbanization pattern of coastal Louisiana can be thought of in general terms as a large urbanized area around New Orleans comprising

Retail Trade	Transportation and Warehousing and Utilities	Information	Finance, Insurance, Real Estate, and Rental and Leasing	Professional, Scientific, Management, Administrative, and Waste Management Services
1,127	442	66	398	373
426	396	52	155	206
3,330	1,221	297	1,357	1,637
25,713	12,595	4,601	14,636	21,668
5,193	2,737	577	1,661	1,725
5,295	1,769	809	2,340	2,935
18,864	11,237	4,596	10,677	18,911
1,051	869	59	409	809
3,632	2,037	468	1,989	2,368
2,278	1,747	312	1,055	1,575
708	581	46	308	405
2,335	1,352	277	996	1,365
2,310	1,355	308	962	1,234
11,423	4,308	2,088	6,045	8,880
5,716	1,842	653	1,960	2,176
5,362	2,780	725	1,728	2,462
2,785	1,291	398	1,190	1,169
97,548	48,559	16,332	47,866	69,898
220,343	98,798	36,418	105,353	140,587
44.3	49.1	44.8	45.4	49.7

continued

Jefferson, Orleans, Plaquemines, St. Bernard, St. Charles, St. James, St. John the Baptist, and St. Tammany Parishes. This metropolitan statistical area, as defined by the U.S. Census Bureau, had a 2000 population of 1.3 million. Small communities are distributed throughout the area wherever people could find land above the wetlands, known as fastland. Other than the metropolitan statistical area, there are only five cities with a population of more than 10,000 in coastal Louisiana: Thibodaux (14,431), Morgan City (12,703), Hammond (17,639), Houma (32,393), and Abbeville (11,887). The 2000 census lists 100 towns with a population of less than 10,000 people.

Although many parishes have zoning and building permits required for development, few have comprehensive plans. The result of having no comprehensive master plan is that when coastal restoration activities (e.g.,

TABLE 3.3 Continued

Parish	Educational, Health, and Social Services	Arts, Entertainment, Recreation, Accommodation, and Food Services	Services
Assumption	1,717	429	329
Cameron	677	269	213
Iberia	4,741	2,079	1,584
Jefferson	41,221	21,705	11,654
Lafourche	7,841	2,012	1,764
Livingston	6,506	2,256	2,224
Orleans	49,315	29,299	10,190
Plaquemines	1,508	812	460
St. Bernard	4,982	2,843	1,445
St. Charles	3,914	1,396	977
St. James	1,527	409	276
St. John the Baptist	2,987	1,461	843
St. Mary	3,409	2,023	1,068
St. Tammany	17,820	7,469	4,360
Tangipahoa	9,796	3,295	2,046
Terrebonne	7,988	3,328	2,193
Vermilion	3,793	1,212	1,032
Total	169,742	82,297	42,658
State totals	402,078	168,593	96,207
Percentage of state totals	42.2	48.8	44.3

SOURCE: Data compiled from U.S. Census Bureau's 2000 census.

river diversions) begin to take place and homes and businesses are relocated, decisions about where people can relocate will have to be made. Without comprehensive planning, it is possible that the homes or businesses affected may move to an area that is being considered as part of a future project or an area facing greater risk as land loss continues. This would result in inefficient use of funds for infrastructure needs, such as potable water, wastewater disposal, utility construction, and roads. There is a need for parish-level, master land-use plans or even a regional land-use plan that includes input from the Louisiana Department of Natural Resources and USACE regarding their plans for coastal restoration.

The Political Will to Coordinate and Collaborate with Proposed Restoration Projects

To date, there appears to be strong support among state and local governments in favor of the projects proposed in the LCA Study and its

TABLE 3.4 Housing Characteristics in Louisiana's Coastal Area

Parish	Total Number of Housing Units	Number of Occupied Housing Units	Percentage of Owner-Occupied Units	Percentage of Rental Units	Percentage of Mobile Homes	Percentage of Housing Units Without Access to a Vehicle
Assumption	9,635	8,239	52.7	15.6	30.4	12.6
Cameron	5,336	3,592	49.3	14.1	29.7	6.6
Iberia	27,844	25,381	53.3	26.4	21.3	10.8
Jefferson	187,907	176,234	58.8	36.1	1.9	9.3
Lafourche	35,045	32,057	58.4	21.7	17.8	9.4
Livingston	36,212	32,603	52.5	16.0	32.0	5.0
Orleans	215,091	188,251	39.6	53.3	0.3	27.3
Plaquemines	10,481	9,021	46.5	20.9	31.5	9.6
St. Bernard	26,790	25,123	65.0	25.3	7.9	10.3
St. Charles	17,430	16,422	69.5	18.6	11.0	6.4
St. James	7,605	6,992	64.1	14.2	20.9	10.2
St. John the Baptist	15,532	14,283	69.7	19.0	12.6	9.5
St. Mary	21,650	19,317	52.5	26.0	21.4	13.2
St. Tammany	75,398	69,253	66.2	19.3	11.4	4.4
Tangipahoa	40,794	36,558	46.4	26.0	23.9	10.3
Terrebonne	39,928	35,997	56.6	24.2	17.6	9.2
Vermilion	22,461	19,832	54.5	22.7	20.4	9.4

SOURCE: Data compiled from U.S. Census Bureau's 2000 census.

predecessors. Further, these entities have demonstrated their understanding of the seriousness of the land loss problem and the need for action.
Active grassroots interaction and careful project selection will be required
to maintain the present solid governmental will to support wetland restoration. The nature of the political system is such that it will depend on
support at the local level. Thus, project design and selection must consider the human element effectively. Early, continuous, and effective
stakeholder engagement will be necessary, and project planners will have
to recognize that maintenance of the present political will require mini-
mal impact on stakeholders and maximum benefits of wetland restoration.

There are numerous challenges to local government with regard to
coastal restoration. The first is overall planning and relocating residences,
businesses, and infrastructure to adapt to coastal restoration projects.
Funding has been allocated for relocation, and this will help; however, it
does not address the problem of how and where to relocate people. A
second challenge will be adapting to the ongoing loss of coastal area. It
may be that parish governments will simply have to acknowledge that
certain areas that continue to subside will have to be abandoned and that
they should focus their attention on areas that can be saved. How this
decision is made will severely test the political system.

Other Broad Challenges and Opportunities

All of the LCA Study's planned restoration efforts to reverse the loss
of coastal lands are proposed to occur in Louisiana. Sound science focuses
on taking a watershed approach to solving water quality and quantity
problems (National Research Council, 2004a). Considering the watershed
(Figure 3.2), the Mississippi River Basin covers 3.27 million km^2 (1.26 million mi^2), or 41 percent of the continental United States. The river itself is
more than 3,540 km (2,200 mi) long, and it has more than 50 navigable
tributaries that comprise about 24,140 km (15,000 mi) of navigable streams
(and thousands of miles of unnavigable ones) (U.S. Army Corps of Engineers, 2001). Activities upstream may be of consequence to the land loss
in Louisiana and the hypoxic zone in the Gulf of Mexico.

A combination of "developments, both natural and man-made, have
occurred that have affected the sediment discharge of the river. The developments include earthquakes; enlargement and closure of distribu-
taries; land use changes; channel dredging; sand and gravel mining; and
the construction of dams, levees, revetments, dikes, and cutoffs" (Kesel,
1988). Human-built structures have been causal factors influencing the
sediment regime of the river since the 1800s. While soil retention and re-

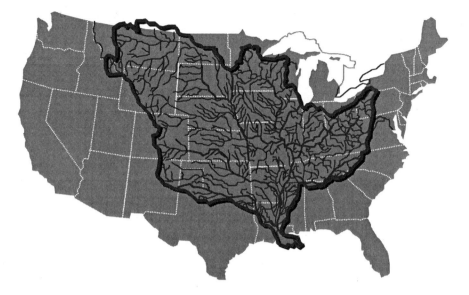

FIGURE 3.2 Mississippi River drainage basin (used with permission from the U.S. Army Corps of Engineers).

duction of bank erosion are considered positive by the people and industries upstream of Louisiana, the reduction of available sediment flowing down the Mississippi River directly impacts the land-building and sustenance processes in the deltaic plain. Kesel (1988) found that there has been a decrease in excess of 80 percent in the amount of suspended sediment transported by the lower Mississippi River below Tarbert Landing, Mississippi, from 1851 to 1982.

The upstream landscape was altered by the destruction of riparian forests and wetlands and by greatly increasing drainage efficiency using ditches, buried drainage tiles, and culverts. Wetlands, reduced by 80 percent in the states of Ohio, Indiana, Illinois, and Iowa, were the critical systems that once helped convert fertilizer nitrate into plant matter and atmospheric nitrogen.

INCREASING THE SUCCESS OF THE
LCA STUDY'S IMPLEMENTATION

Achieving a desired state of restoration or level of protection will require balancing a number of natural and anthropogenic processes. Although some of these processes take place over areas much larger than that covered by the LCA Study, most of the more significant processes operate locally. Understanding how these processes change with time and in response to human activity is, and should be, a major focus of the Science and Technology Program envisioned in the LCA Study.

Given the nature of the Mississippi River Delta, it is understandable that any solution approaching a large-scale or optimal restoration will encounter conflicts between navigation, flood control, oil and gas channels, and other social and economic activities on the one hand and the need for large-scale redistribution of Mississippi River freshwater and sediment on the other. This underscores the need for the LCA Study to identify at an early stage those projects that can be undertaken successfully, to convey this information to the general public, and to establish reasonable expectations for accomplishments. Diverting river water to build land and marshes requires breaching levees and flooding areas. The large amount of money—up to 35 percent of a project's total cost—for land purchase and relocation of homes, residences, and businesses reflects the reality that currently inhabited or agriculturally productive lands will be flooded. Gaining public acceptance for wetland restoration will be difficult when homes and businesses require relocation or when owners are prevented from reoccupying properties damaged by recent hurricanes that are in areas needed to support restoration efforts.

The LCA Study does not propose any programs that might stem any upper watershed contributors to coastal land loss and hypoxia in the Gulf of Mexico. None of the plans proposed have attempted to consider contributing factors outside Louisiana for any projects. If the wetlands of concern are deemed of national significance, the range of solutions should encompass the entire nation or, at a minimum, the entire watershed.

As discussed in Chapter 2, land loss in coastal Louisiana is due to a variety of natural and human-related causes; the role of each causal factor varies with location and time and is not well quantified. High among the causes are relative sea level rise (including high subsidence rates) across broad areas; growth faults; access canals and navigation waterways; channelization of the Mississippi River by levees, causing much of the river-borne sediments to be conveyed to deep water at the terminus of the Birdsfoot Delta; grazing by fur-bearing animals; and processes causing wetlands to erode in some areas and build in others. Some of these individual causative factors encompass both natural and anthropogenic elements. The established interests of maintaining a levee system in which sediment deposition in channels and to adjacent lowlands is reduced are counter to the delivery of freshwater, sediments, and nutrients to areas in need of wetland restoration and maintenance. Restoring the essential and widespread distribution of sediment and freshwater flow, while maintaining stakeholder acceptance of the adverse impacts, will be the overarching challenge to future comprehensive efforts to realize the vision espoused by Coast 2050.

The natural and anthropogenic processes contributing to net land loss in coastal Louisiana are significant and pervasive and have been operating for decades. Furthermore, because the sediment supply is limited, the affected area is large, and the social, political, and economic impediments are extensive, achieving no net loss may be problematic. By its own analysis, the LCA Study points out that implementing the restoration efforts it proposes would reduce land loss by about 20 percent (from 26.7 km^2 per yr [10.3 mi^2 per yr] to 22.3 km^2 per yr [8.6 mi^2 per yr]) at a cost of approximately \$18 million per km^2 per yr (\$47 million per mi^2 per yr). Thus, future efforts will have to be more elaborate than those proposed in the LCA Study, or expectations will have to be reduced.

Efforts to restore significant portions of coastal Louisiana would entail changing the current geographic distribution of land, water, and wetland. Land use and infrastructure development (e.g., roads, pipelines, utilities) have changed in response to the changing coastline. The proposed projects will again force change in the way people work, live, and play in the area. One way to deal efficiently with the change is through comprehensive land-use planning that is coordinated with the planned restoration projects.

A survey of local parish governments indicates that all but one of the parishes has a planning department. Only 10 of the 20 parish governments, however, have a comprehensive land-use plan, and at least four of the plans are more than 10 years old. **The parishes should develop comprehensive land-use plans in order for there to be orderly and economically efficient relocation of infrastructure, homes, and businesses during coastal restoration (as planned for in the LCA Study).** Clearly, effective land-use plans that act in concert with and support a comprehensive restoration effort will require a widely understood and accepted "end state" of restoration efforts.

A number of sociopolitical challenges involving the various stakeholder interests discussed in this chapter will place limitations on what can be achieved through any restoration effort. **Louisiana's coastal restoration plans must acknowledge these limitations prominently and adjust goals and public expectations accordingly.** With the certainty that even the most optimistic Louisiana coastal landform of the future will differ from that at present, the emphasis should be on establishing realistic estimates of future landforms and conveying these to stakeholders. **Restoration efforts should be focused to maximize targeted ecological, social, and economic benefits while promoting managed retreat in selected regions.** This could involve reducing the rate of land loss in key areas and allowing the system to approach natural equilibrium in others.

Future efforts must be more realistic in considering the location patterns of human settlements relative to project locations, including the option of infrastructure depreciation and abandonment. These sociopolitical challenges will be revisited in subsequent chapters as the adaptive management mechanisms currently included in the LCA Study are reviewed.

4

Plans and Efforts at Restoring Coastal Louisiana

HIGHLIGHTS

This chapter
  • Reviews the extensive planning, development, and restoration efforts that have preceded and laid the foundation for the *Louisiana Coastal Area (LCA), Louisiana—Ecosystem Restoration Study* (LCA Study)
  • Identifies the elements of previous efforts that would be desirable components of the LCA Study under review
  • Reviews goals and objectives of ongoing and planned activities outside of the LCA Study

The previous chapters suggest some fundamental axioms regarding the Louisiana coastal zone:

1. Land loss in coastal Louisiana presents a complex scientific and social problem of unprecedented scale that will require a multi-decadal management solution. A clear, realistic, and shared vision of restoration goals for the Louisiana coast is critical to the successful implementation of the LCA Study.

2.   Some Louisiana residents will be impacted by changes in this very productive ecosystem whether or not the actions discussed in the LCA Study are undertaken.

3.   There are management and engineering actions that could be done to improve the current condition in some areas.

4.   Land loss is a very urgent problem that with time will only become worse and more costly to address.

5.   It is not possible to restore the entire area, and some communities and existing habitats will be lost.

6.   Large parts of the ecosystem will undergo shifting habitat types, and the rate of change and ability of the ecosystem to adapt will depend on the management strategy adopted.

As stated in numerous U.S. Army Corps of Engineers (USACE) policy statements and recommended in many recent National Research Council reports, planning and implementation of water resources projects (including those involving environmental restoration) should be undertaken within the context of the larger system. In keeping with this system approach, comprehensive efforts to address land loss in coastal Louisiana should treat the entire area as an integrated system (actions taken in one locale will have a direct or indirect influence elsewhere in the Mississippi River Delta). The overall project management approach should also be transparent, scientifically based and defensible and be implemented through an adaptive management process to adjust to the best available knowledge of the region. Both the urgency and the need for comprehensive planning initiatives have been identified by earlier independent reviews (Boesch et al., 1994). Efforts to restore coastal Louisiana started about 30 years ago, and this chapter chronicles the more recent efforts. (See Box 1.1 for a list of efforts from 1967 to the release of the LCA Study.)

COASTAL WETLANDS PLANNING, PROTECTION, AND RESTORATION ACT

Legislative History and Funding

In recognition of the national importance of wetlands and the significant wetland losses occurring in Louisiana and elsewhere, Congress passed the Coastal Wetlands Planning, Protection, and Restoration Act (CWPPRA) (P.L. 101-646, Title III) in 1990. CWPPRA establishes both a national wetlands conservation initiative and a wetlands conservation and restoration program for the State of Louisiana. The objectives of CWPPRA are as follows:

• Provide for the planning, identification, and implementation of priority coastal wetland restoration projects in Louisiana

• Encourage the State of Louisiana to develop a plan with the goal of achieving no net loss of coastal wetlands as a result of future development activities

• Provide for grants to coastal states to implement coastal wetlands conservation projects

Funding authority for CWPPRA is provided through the Budget Reconciliation Act of 1990. Funds are derived from a portion of the Aquatic Resources Trust Fund that is supported by taxes on fishing equipment and motorboat and engine fuel tax. Of the amount appropriated each year from this fund, the Department of the Interior receives 30 percent to support the National Coastal Wetlands Conservation Grant Program and the North American Wetlands Conservation Fund. USACE receives 70 percent for construction and associated activities related to the Louisiana program. Funding for the Louisiana program is approximately $50 million per year. The cost share for CWPPRA projects is 85 percent by the federal government and 15 percent by the State of Louisiana.

Organization

CWPPRA directed the Secretary of the Army to establish a task force composed of the Administrator of the Environmental Protection Agency, the Secretary of the Department of Commerce, the Secretary of the Department of the Interior, the Secretary of the Department of Agriculture, and the Governor of the State of Louisiana. Senior officials, who have been delegated responsibility, represent the official members of the task force. The New Orleans District Commander of USACE is the task force chairman. The New Orleans District also provides general administrative and management support to the task force and has financial accounting and disbursement responsibility for all federal and nonfederal funds associated with the program.

The task force is the overall governing and final decision-making body of the CWPPRA program in Louisiana (Figure 4.1). The task force prepares and submits to Congress an annual project priority list (PPL) of Louisiana wetland restoration projects.

The technical committee and planning and evaluation subcommittee each have the same representation as the task force. The technical committee provides advice and recommendations to the task force on the engineering, environmental, economic, real estate, construction, operation and maintenance, and monitoring aspects of the CWPPRA program and individual projects. The planning and evaluation subcommittee is the day-

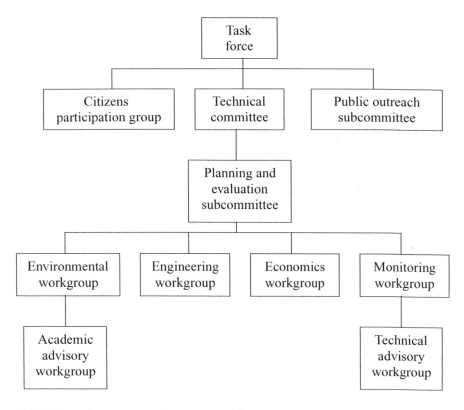

FIGURE 4.1 Organizational structure of the CWPPRA program (U.S. Geological Survey, 2005; used with permission from the U.S. Geological Survey).

to-day working level committee that oversees technical workgroups and provides input and recommendations to the technical committee.

The planning and evaluation subcommittee established four technical groups for environment, engineering, economics, and monitoring to evaluate projects for PPL and restoration plan. The environmental workgroup estimates the benefits (in terms of wetlands created, protected, enhanced, or restored) associated with proposed projects. The engineering workgroup reviews project cost estimates for consistency. The economic workgroup performs the economic analysis that permits comparison of projects based on cost-effectiveness. The monitoring workgroup established a standard procedure for monitoring CWPPRA projects, developed a monitoring cost estimating procedure based on project type, and provides advice and recommendations on all CWPPRA monitoring activities. The technical advisory workgroup reports to the monitoring workgroup and assists with the design of project-specific monitoring plans and conducts evaluations of project effectiveness.

An academic advisory workgroup represents Louisiana's academic community and provides general support for the program, including the screening, development, and ranking of demonstration projects, as well as assisting with individual project design. This group reports to the environmental workgroup.

To facilitate public input into the process, a public outreach subcommittee was established by the task force to oversee and conduct the communications, public relations, and education activities of the program. Participating federal agencies, the State of Louisiana, and nonprofit organizations make up the membership. There is also a citizen's participation group to receive input from the public and to promote public participation in the program. This group assists with developing PPLs and ensures that the general public has an opportunity to review PPLs.

Accomplishments

The CWPPRA task force developed the required Louisiana Coastal Wetlands Restoration Plan in 1993. This plan provided a basis for early PPL project selection. The Louisiana Coastal Wetlands Restoration Plan divided the Louisiana coastal zone into nine hydrologic basins. Basin-level restoration solutions using proven techniques with cost and benefit analyses were developed to respond to the specific priority needs of each basin. The plan proposed projects estimated at $1.3 billion that could prevent about 65 percent of coastal land losses over 20 years (U.S. Army Corps of Engineers, 2005a).

In response to the CWPPRA legislation, the Louisiana Department of Natural Resources was designated the lead agency to develop a state wetlands conservation plan. The conservation plan included regulatory, nonregulatory, and educational programs that would achieve the desired no net loss goal (Louisiana Coastal Wetlands Conservation and Restoration Task Force and the Wetlands Conservation and Restoration Authority, 1998). Approval of the wetlands conservation plan in 1997 resulted in a revision of the cost-sharing requirements of CWPPRA from the 75 percent to the current 85 percent federal share.

Restoration Projects

Since passage of CWPPRA in 1990, the task force has prepared and submitted 13 PPLs to Congress. CWPPRA projects fall into 10 categories of restoration techniques:

1. Freshwater diversion
2. Freshwater outfall management
3. Sediment diversion
4. Dredged material or marsh creation
5. Shoreline protection
6. Sediment and nutrient trapping
7. Hydrologic restoration
8. Marsh management
9. Barrier island restoration
10. Vegetation planting

The annual selection of projects begins with input and recommendations from the public and task force member agencies, taking into consideration strategies outlined in *Coast 2050: Toward a Sustainable Coastal Louisiana* (Coast 2050). (See below for more details.) Reviews and evaluations are conducted by the task force workgroups, and final recommendations for project selection for PPLs are based on the proposed project's technical (scientific) merit, cost-effectiveness, and predicted benefits in terms of improving wetland quantity and quality. The predicted benefits of potential projects are determined by using Wetland Value Assessment models. These models were developed specially for CWPPRA in order to provide a quantitative means to compare expected changes in habitat quantity and quality across projects (Louisiana Coastal Wetlands Conservation and Restoration Task Force, 2003).

Each project included in a PPL has a project life of approximately 20 years, during which time the project is maintained and monitored to determine the long-term effectiveness of the restoration effort. Since 1991, 142 projects have been approved with a projected benefit to create, restore, or protect almost 566 square kilometers (km²) (218.5 square miles [mi²]) of coastal wetlands over the next 20–30 years at an estimated cost of $504 million (Louisiana Coastal Wetlands Conservation and Restoration Task Force, 2003; Table 4.1). There are currently 124 active CWPPRA projects. It is estimated that the total land gain due to the CWPPRA projects over the next 50 years will be 140 km² (54 mi²) (U.S. Army Corps of Engineers, 2004a). In 1997, the projected prevention of total wetland losses accrued by CWPPRA was revised to be less than 25 percent of the projected land lost by 2050 (U.S. Army Corps of Engineers, 2005a). This led to recognition that a new, more robust restoration effort was needed. As a result, in 1998, the Coastal Wetlands Conservation and Restoration Task Force and the Wetlands Conservation and Restoration Authority prepared Coast 2050.

TABLE 4.1 Projected Results of Coastal Wetlands Planning, Protection, and Restoration Act (CWPPRA) Initiatives (as of 11/03)

	CWPPRA Projects Authorized	CWPPRA Projects Constructed	Expected Wetland Benefits in km^2 (mi^2 in parentheses) (authorized projects)[a]	Total Cost Estimates (authorized projects) (dollars)
Region 1 Basins	17	7	*48.0 (18.5)*	*25,475,934*
Pontchartrain	17	7	48.0 (18.5)	25,475,934
Region 2 Basins	42	14	*267.1 (103.1)*	*171,267,903*
Breton Sound	6	1	14.3 (5.5)	10,742,032
Mississippi River	9	4	174.2 (67.3)	33,082,303
Barataria	27	9	78.6 (30.3)	127,443,568
Region 3 Basins	47	23	*82.2 (31.7)*	*191,813,675*
Terrebonne	32	14	37.6 (14.5)	150,471,719
Atchafalaya	3	2	17.7 (6.8)	11,965,718
Teche/Vermilion	12	7	26.9 (10.4)	29,376,238
Region 4 Basins	33	22	*108.2 (41.8)*	*99,246,747*
Mermentau	14	7	25.8 (10.0)	27,369,914
Calcasieu/Sabine	19	15	82.4 (31.8)	71,876,833
Coastwide	3	2	*60.6 (23.4)*	*16,233,889*
Total	*142*	*68*	*566.1 (218.6)*	*504,038,148*

[a]Expected wetland benefits are defined as the number of wetland acres created, restored, or protected over the 20-year project life.

SOURCE: Modified from Louisiana Coastal Wetlands Conservation and Restoration Task Force, 2003; used with permission from the Louisiana Department of Natural Resources.

COAST 2050

Coast 2050 has been described as a strategic plan for the survival of Louisiana's coast. It has also been called an "unprecedented effort among diverse groups who have united behind a common vision to sustain a coastal ecosystem that supports and protects the environment, economy, and culture of southern Louisiana and that contributes greatly to the economy and well-being of the nation" (Louisiana Coastal Wetlands Conservation and Restoration Task Force and the Wetlands Conservation and Restoration Authority, 1998).

Coast 2050 was a joint coastal restoration planning effort undertaken by federal, state, and local entities as well as academics and other interested parties. The planning effort sought to maximize common ground between ecosystem needs and publicly acceptable restoration solutions (Figure 4.2). "The process involved an integrated, multiple-use approach to ecosystem management and considered such factors as fish and wild-life productivity, transportation, navigation, utilities infrastructure, fresh-water supply, public safety, local economies, businesses, jobs, and community stability" (Louisiana Coastal Wetlands Conservation and Restoration Task Force and the Wetlands Conservation and Restoration Authority, 1998). The Louisiana coast was divided into four regions representing distinct geologic and hydrologic areas (Figure 1.1). Each region was used as a basis for analysis and to facilitate local input into the plan

Figure 4.3 illustrates the multiagency and federal and state partnership that evolved in developing Coast 2050. Represented in the strategic working group were USACE, the Environmental Protection Agency (EPA), the Natural Resources Conservation Service (NRCS), the U.S. Fish and Wildlife Service (FWS), and the National Oceanic and Atmospheric Administration (NOAA) Fisheries. Many state agencies were represented on the strategic working group, which also included academic and consultancy support. This working group was responsible for overseeing

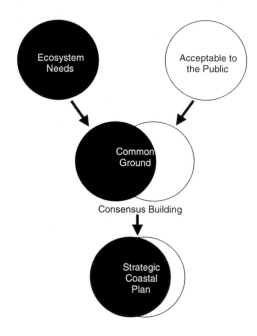

FIGURE 4.2 Coast 2050 process for public involvement and strategic planning (Louisiana Coastal Wetlands Conservation and Restoration Task Force and the Wetlands Conservation and Restoration Authority, 1998; used with permission from the Louisiana Department of Natural Resources).

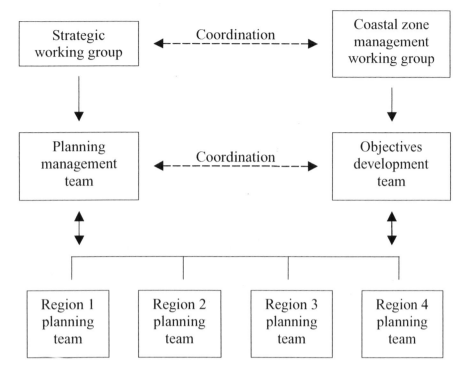

FIGURE 4.3 Coast 2050 organization (Louisiana Coastal Wetlands Conservation and Restoration Task Force and the Wetlands Conservation and Restoration Authority, 1998; used with permission from the Louisiana Department of Natural Resources).

the development of the strategic plan. The planning management team was responsible for authoring Coast 2050.

The coastal zone management working group consisted of parish government representatives and parish coastal zone management advisory committees. The coastal zone management working group was responsible for determining public acceptance of habitat objectives and restoration strategies. To that end, the public participated in 65 public meetings held throughout the planning process from 1997 to 1998 throughout the study area. The objectives development team focused on obtaining information regarding coastal use and resource objectives that were used to create the strategic plan.

Four regional planning teams were established to create coastal restoration strategies and to review the coastal-use and resource objectives generated by the objectives development team. These planning teams, which comprised agency staff, academics, parish governments, Louisiana Cooperative Extension Service-Louisiana State University Sea Grant staff, and volunteer local participants, provided technical information and proposed regional coastal strategies to the planning management team.

The CWPPRA task force, the State Wetlands Authority, and the Louisiana Department of Natural Resources' Coastal Zone Management Authority adopted Coast 2050 as their unified coastal restoration strategy. All CWPPRA projects beginning in 1999 were required to reflect the strategies outlined in Coast 2050. All 20 coastal parish governments adopted resolutions in support of Coast 2050 (Louisiana Coastal Wetlands Conservation and Restoration Task Force and the Wetlands Conservation and Restoration Authority, 1998).

Coast 2050 was an extension of work conducted in previous state coastal planning efforts. Results from CWPPRA projects were also folded into the analysis. New technical information was developed (e.g., projecting wetland losses between 1990 and 2050) considering faulting, subsidence, and land loss in coastal Louisiana. Methods were developed to assess existing trends in fisheries production and to project these into the future. Existing wildlife habitat status and future trends were analyzed (Louisiana Coastal Wetlands Conservation and Restoration Task Force and the Wetlands Conservation and Restoration Authority, 1998).

Coast 2050 focuses on processes, such as river water diversions, in order to create and sustain marsh by accumulating sediment and organic matter and to maintain habitat diversity by varying salinities and protecting key landforms. Dredged materials would be used beneficially to create marsh in various sites. Barrier islands, headlands, and shorelands would be restored and maintained using the most cost-effective means. The Mississippi River Gulf Outlet (MRGO) navigation channel was to be closed as soon as possible. In the Barataria-Terrebonne area, large amounts of water from the Mississippi River would be funneled to build two new deltas, one on either side of Bayou Lafourche. The Atchafalaya River would continue to carry muddy sediment east and south to support nearby marshes. In the Calcasieu-Sabine area, seasonally operated locks at the mouth of the navigation channels would help the marshes recover from salinity stress (Louisiana Coastal Wetlands Conservation and Restoration Task Force and the Wetlands Conservation and Restoration Authority, 1998).

RECONNAISSANCE-LEVEL REPORT

The next step undertaken by USACE was the development and release of the 905(b) reconnaissance report (U.S. Army Corps of Engineers, 1999a), which was required by USACE's water resources planning process. This report recommended that Coast 2050 proceed to the feasibility phase, which would produce the *Louisiana Coastal Area, LA—Ecosystem Restoration: Comprehensive Coastwide Ecosystem Restoration Study* (draft LCA Comprehensive Study) that assesses Coast 2050 at a programmatic, feasibility level. Along with the draft LCA Comprehensive Study, a programmatic environmental impact statement (PEIS) was developed starting with scoping meetings in April 2002 (U.S. Army Corps of Engineers, 2004b). PEIS analyzes project features that are non-location specific, provides a macro description of economic effects, and gives parametric cost estimates.

DRAFT LCA COMPREHENSIVE STUDY

Begun as a substitute for a series of feasibility reports, the draft LCA Comprehensive Study was based on Coast 2050 and the 905(b) reconnaissance report. It was intended to serve as a national ecosystem restoration plan and to achieve a coastwide system of restoration projects in Louisiana that would function in an integrated fashion to achieve established goals. The draft LCA Comprehensive Study, made available by the State of Louisiana in November 2003, describes and summarizes a series of studies that was to be reviewed and adjusted based on public and technical review.[1] As discussed later, even though a formal comprehensive plan has not been released, much of the work completed and discussed in the draft LCA Comprehensive Study provides the underpinnings of the LCA Study. Thus, its discussion here is appropriate. The draft LCA Comprehensive Study contains a set of alternative comprehensive plans for each of the four regions but offers no specific recommended plan. (While a comprehensive plan is needed, this does not necessarily imply endorsement of *the* draft LCA Comprehensive Study.)

Goal and Objectives

The draft LCA Comprehensive Study (U.S. Army Corps of Engineers, 2003a) sought to provide the following objectives:

- Create a sustainable Louisiana coastal ecosystem having the essential functions and values of a natural ecosystem
- Restore the largest practicable acreage of productive and diverse wetlands

---

[1]The draft LCA Comprehensive Study, like the LCA Study itself, would have been followed up by a series of decision documents (feasibility reports) evaluating, in detail, certain components of the overall restoration program (G. Duszynski, written communication, 2005).

- Develop restoration plans in consideration of the economic impacts and incidental economic benefits to natural resources, communities, infrastructure, and industries along the coast
- Develop a comprehensive plan that is coordinated and consistent with other major land-use and infrastructure features, particularly with respect to navigation, hurricane protection or flood control, and oil and gas production

Organization

An interagency project delivery team[2] was assembled with staff from USACE, the State of Louisiana, FWS, NOAA, EPA, the U.S. Geological Survey (USGS), and the U.S. Department of Agriculture (USDA). Further, USACE and the State of Louisiana assembled a team of more than 120 scientists, engineers, and planners from across the nation to provide advice and guidance, carry out complex modeling efforts, and review results (U.S. Army Corps of Engineers, 2003a; Figure 4.4).

As part of this broad effort, USACE established the National Technical Review Committee (NTRC) in April 2002 to provide an external, independent technical review of the draft LCA Comprehensive Study. NTRC consists of 10 internationally recognized scientists encompassing a range of technical disciplines. Seven meetings were held in New Orleans between June 2002 and November 2003 to provide reviews at strategic times during the development of the draft LCA Comprehensive Study. NTRC members provided individual comments and recommendations after each meeting, and a report was prepared (U.S. Army Corps of Engineers, 2003a; Orth et al., 2005).

Planning and the Plan Development Process

The draft LCA Comprehensive Study's planning process consisted of the following five phases:

- Phase I: Establish planning objectives and evaluation criteria
- Phase II: Assess restoration strategies from Coast 2050
- Phase III: Develop and evaluate restoration projects and measures
- Phase IV: Develop and evaluate alternatives—Select a final array of coastwide plans
- Phase V: Select coastwide plan that best meets objectives (to be accomplished after public coordination)

---

[2]The project development team is the formal mechanism described in the National Environmental Policy Act.

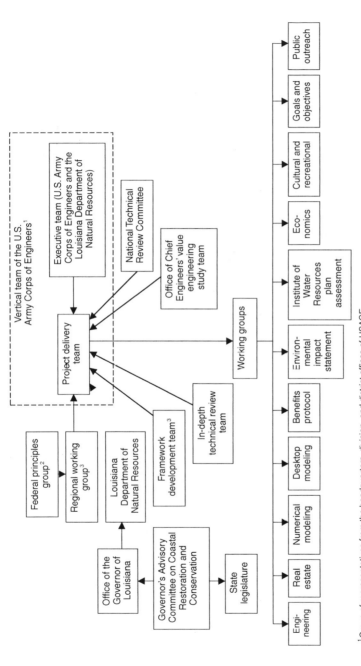

FIGURE 4.4 Plan formulation for the draft LCA Comprehensive Study. (Information was gathered from the U.S. Army Corps of Engineers [2003a].)

[1] Group of representatives from the headquarters, division, and district offices of USACE.
[2] EPA, FWS, the Minerals Management Service, NOAA Fisheries, USGS, NRCS, the U.S. Department of Energy, the U.S. Department of Transportation, and the Federal Emergency Management Agency.
[3] Federal interagency representatives, state representatives, and environmental nongovernmental organizations.

Phases I and II built on previous efforts so the most elaborate and detailed work took place in Phases III and IV. In Phase III, projects were developed for each region and evaluated for contributions to three increasingly ambitious quantitative target levels of restoration: (1) reduce the present rate of total wetland loss by 50 percent of the no-action land loss rate, (2) maintain the present total area of coastal wetlands over the next 50 years, and (3) enhance the present landscape condition by increasing the wetland area by 50 percent of the present annual net land loss rate over the next 50 years. Because not all of the proposed projects can be combined and interactions among project effects must be anticipated, the hundreds of potential projects were grouped into 32 alternative region plans: 10 each for the Pontchartrain–lower Mississippi–Breton Sound and Barataria regions, five for the Teche–Vermilion–Atchafalaya–Terrebonne region, and seven in the Chenier Plain region.

The draft LCA Comprehensive Study's planning process was guided by three hydrogeomorphic objectives and two ecosystem objectives (Box 4.1). The process involved subjecting the various restoration alternatives to a fairly straightforward scale of potential effectiveness. In addition, seven environmental operating principles (Box 4.2) informed the formu-

---

Box 4.1
Tactical Planning Objectives

Hydrogeomorphic Objectives:

• Establish dynamic salinity gradients that reflect natural cycles of freshwater availability and marine forcing (fluctuation related to normal daily and seasonal tidal action and exchange)
• Increase sediment input from sources outside estuarine basins and manage existing sediment resources within estuarine basins to sustain and rejuvenate existing wetlands and rebuild marsh substrate
• Maintain or establish natural landscape features and hydrologic processes that are critical to sustainable ecosystem structure and function

Ecosystem Objectives:

• Sustain productive and diverse fish and wildlife habitats
• Reduce nutrient delivery to the continental shelf by routing Mississippi River waters through estuarine basins while minimizing potential adverse effects

SOURCE: U.S. Army Corps of Engineers, 2003a.

Box 4.2
Environmental Operating Principles

1. Strive to achieve environmental sustainability and recognize that an environment maintained in a healthy, diverse, and sustainable condition is necessary to support life
2. Recognize the interdependence of life and the physical environment and proactively consider environmental consequences of USACE programs and act accordingly in all appropriate circumstances
3. Seek balance and synergy among human development activities and natural systems by designing economic and environmental solutions that support and reinforce one another
4. Continue to accept corporate responsibility and accountability under the law for activities and decisions under our control that impact human health and welfare and the continued viability of natural systems
5. Seek ways and means to assess and mitigate cumulative impacts to the environment and bring systems approaches to the full life cycle of our processes and work
6. Build and share an integrated scientific, economic, and social knowledge base that supports a greater understanding of the environment and impacts of our work
7. Respect the views of individuals and groups interested in USACE activities, listen to them actively, and learn from their perspective in the search to find innovative win-win solutions to the nation's problems that also protect and enhance the environment

SOURCE: U.S. Army Corps of Engineers, 2004a.

---

lation process and are integrated into all proposed program and project management processes (U.S. Army Corps of Engineers, 2004a). The project delivery team compiled 10 guiding principles for plan formulation (Box 4.3) in coordination with key stakeholder groups and with public comments provided during the scoping process (U.S. Army Corps of Engineers, 2004a).

In Phase IV, the likely outcomes of each regional alternative were assessed using simulation models, desktop models, and restoration benefit calculations. (See Chapter 5 for a description and analysis of these models.) Here, the term "benefit" is used to signify generic contribution to the objectives (e.g., acres of reduced land loss or of land gain), not assessment of the economic benefits. In an extensive cost-effectiveness and incremental cost analysis, the analysts compared benefits per unit cost among the

---

Box 4.3
Ten Guiding Principles for Plan Formulation

1. It is evident that management of Louisiana's coast is at a point of decision. Only a concerted effort now will stem this ongoing degradation, and thus alternatives must include features which can be implemented in the near-term and provide some immediate benefits to the ecosystem, as well as those which require further development and refinement of techniques and approaches.

2. Appreciation of the natural dynamism of the coastal system must be integral to planning and the selection of preferred alternatives. This should include assessing the risks associated with tropical storms, river floods, and droughts.

3. Alternatives that mimic natural processes and rely on natural cycles and processes for their operation and maintenance will be preferred.

4. Limited sediment availability is one of the constraints on system rehabilitation. Therefore, plan elements including mechanical sediment retrieval and placement may be considered where landscape objectives cannot be met using natural processes. Because sediment mining can contribute to ecosystem degradation in the source area, such alternatives should, to the extent practicable, maximize use of sediment sources outside the coastal ecosystem (e.g., from the Mississippi River or the Gulf of Mexico).

5. Plans will seek to achieve ecosystem sustainability and diversity while providing interchange and linkages among habitats.

6. Future rising sea levels and other global changes must be acknowledged and incorporated into planning and the selection of preferred alternatives.

---

alternative combinations of projects in each region, added some supplemental plans to "address completeness," and selected a final array of seven alternative comprehensive plans for further review. In parallel with the development of the draft LCA Comprehensive Study, USACE drafted a PEIS for review and comment by the public.

Public Outreach and Consensus Building

Based on the draft PEIS, USACE and the State of Louisiana held six scoping meeting in April 2002. They received 301 verbal and written comments that are summarized in the draft PEIS. In addition to the PEIS comment process, the project development team pursued a multi-tiered plan for public involvement, including (1) special meetings with local governments in the parishes; (2) public meetings to solicit input from stakehold-

7. Displacement and dislocation of resources, infrastructure, and possibly communities may be unavoidable under some scenarios. In the course of restoring a sustainable balance to the coastal ecosystem, sensitivity and fairness must be shown to those whose homes, lands, livelihoods, and ways of life may be adversely affected by the implementation of any selected alternatives. Any restoration-induced impacts will be consistent with [the National Environmental Policy Act] in that actions will be taken to avoid, minimize, rectify, reduce, and then, only if necessary, compensate for project-induced impacts.

8. The rehabilitation of the Louisiana coastal ecosystem will be an ongoing and evolving process. The selected plan should include an effective monitoring and evaluation process that reduces scientific uncertainty, assesses the success of the plan, and supports adaptive management of plan implementation.

9. Recognizing that disturbed and degraded ecosystems can be vulnerable to invasive species, implementation needs to be coordinated with other state and federal programs addressing such invasions, and project designs will promote conditions conducive to native species by incorporating features, where appropriate, to protect against invasion to the extent possible without diminishing project effectiveness.

10. Net nutrient uptake within the coastal ecosystem is maximized through increased residence time and the development of organic substrates, and thus project design should promote conditions that route riverine waters through estuarine basins and minimize nutrient export to shelf waters.

SOURCE: U.S. Army Corps of Engineers, 2004a.

ers, other governmental units, and academia; (3) Internet web site interactions; and (4) special briefings for executives of large corporations and national interest groups. To obtain public input to the process, the project delivery team held four public meetings during February 2003 regarding plan formulation, four meetings in May–June 2003 to present and receive comments on the 32 region alternatives, and four meetings during August 2003 to present and receive comments on the final array of alternatives being considered. The November 2003 draft PEIS contains a summary of the comments received at the public meetings.

LCA STUDY

Due to the Office of Management and Budget's desire to shift the focus to a smaller, near-term effort, USACE prepared the LCA Study. Some of the analysis completed for the draft LCA Comprehensive Study was used to support the LCA Study's conclusions regarding mechanisms for attaining near-term restoration goals. (Refer to Chapter 1 for additional information.) In short, the goals of the LCA Study are to (1) identify critical human and ecological needs, (2) determine alternatives for meeting these needs, (3) identify restoration features where construction can begin within 5–10 years, (4) establish priorities within these restoration features, (5) describe how the priority features could be developed and implemented, (6) determine the scientific uncertainties and engineering challenges and propose solutions to these uncertainties and challenges, (7) identify feasibility studies that need to be done within the next 5–10 years to explore large-scale, long-term efforts, and (8) develop a strategy to address the long-term restoration needs of coastal Louisiana.

IMPLEMENTATION OF THE LCA STUDY: ORGANIZATION, DURATION, AND FUNDING

The implementation of the LCA Study is expected to be organized as depicted in Figure 4.5. The LCA Study requests "specific Congressional authorization for five near-term critical restoration features for which construction starts in 5–10 years, with implementation subject to approval of feasibility-level documents by the Secretary of the Army" (U.S. Army Corps of Engineers, 2004a). The total project costs are summarized in Table 4.2. The portion of the LCA Study for which immediate authorization is being requested includes the five near-term critical projects ($864,065,000), the Science and Technology (S&T) Program ($100,000,000), the demonstration project program ($100,000,000), the beneficial use program ($100,000,000), and the investigations of modifications of existing structures ($10,000,000) at a total cost of $1,174,065,000. Construction authorization was requested from Congress for the five critical projects because significant feasibility-level work had already been performed on them, and waiting for construction authorization in future Water Resources Development Act (WRDA) bills could delay their construction. If authorization is provided, decision documents (individual feasibility reports) would still be required prior to construction, but the decision to proceed to construction would be made by the Assistant Secretary of the Army for Civil Works rather than Congress (G. Duszynski, written communication, 2005).

Other components of the LCA Study include 10 additional near-term critical projects that will require future congressional authorization for construction ($761,916,000) and studies of six potentially promising long-term, large-scale restoration concepts ($60,000,000). Thus, the estimated cost of implementing all components of the LCA Study is $1,995,981,000;

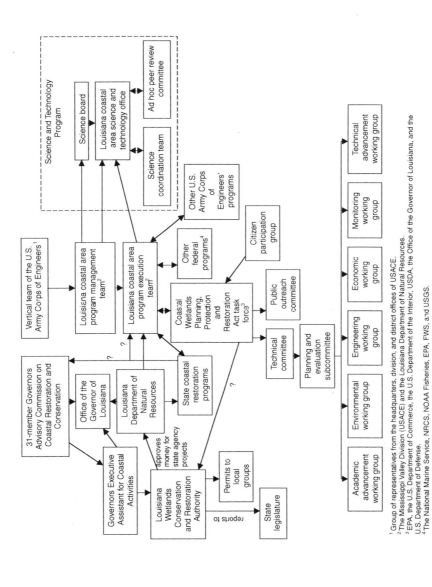

FIGURE 4.5 LCA Study implementation plan. (Information was gathered from the U.S. Army Corps of Engineers [2004a].)

[1] Group of representatives from the headquarters, division, and district offices of USACE.
[2] The Mississippi Valley Division (USACE) and the Louisiana Department of Natural Resources.
[3] EPA, the U.S. Department of Commerce, the U.S. Department of the Interior, USDA, the Office of the Governor of Louisiana, and the U.S. Department of Defense.
[4] The National Marine Service, NRCS, NOAA Fisheries, EPA, FWS, a nd USGS.

TABLE 4.2 Proposed Funding (Dollars) for the LCA Study with Federal/State Cost Share Ratios

Item	Federal Share	State Share	Total Cost
Feasibility-level decision and National Environmental Policy Act documentation (50/50)	27,336,500	27,336,500	54,673,000
Near-term feature first construction cost (65/35)	402,750,300	61,758,450	489,845,000
Preconstruction, engineering, and design (65/35)	24,327,500	11,924,500	36,252,000
Engineering and design (65/35)	20,277,850	8,740,150	29,018,000
Supervision and administration (65/35)	48,859,750	20,113,250	68,973,000
Project monitoring (65/35)	4,553,150	2,131,850	6,685,000
Land, easements, rights of way, relocation, and disposal (LERRD) (0/100)	0	178,619,000	178,619,000
Conditionally authorized[a] subtotal	**528,105,050**	**310,623,700**	**864,065,000**
Cash contribution	*555,225,300[b]*	*130,004,450*	
Science and Technology Program (10 years) (65/35)	65,000,000	35,000,000	100,000,000
Demonstration program (10 years) (65/35)	65,000,000	35,000,000	100,000,000
Beneficial use of dredge material program (72/25)	75,000,000	25,000,000	100,000,000
Investigations of modifications of existing structures (50/50)	5,000,000	5,000,000	10,000,000
Programmatically authorized[c] subtotal	**210,000,000**	**100,000,000**	**310,000,000**
Cash contributions	*210,000,000*	*100,000,000*	
Total conditionally and programmatically authorized subtotal	**738,105,050**	**410,623,700**	**1,174,065,000**
Feasibility-level decision and National Environmental Policy Act documentation (50/50)	23,764,500	23,764,500	47,529,000
Near-term feature first construction cost (65/35)	334,439,850	25,829,150	360,269,000
Preconstruction, engineering, and design (65/35)	31,417,550	4,609,450	36,027,000
Engineering and design (65/35)	40,662,750	4,972,250	45,635,000

TABLE 4.2 Continued

Item	Federal Share	State Share	Total Cost
Supervision and administration (65/35)	54,137,450	4,535,550	58,673,000
Project monitoring (65/35)	3,693,950	1,989,050	5,683,000
LERRD (0/100)	0	208,100,000	208,100,000
Conventionally authorized[d] features subtotal	**488,116,050**	**273,799,950**	**761,916,000**
Cash contributions	*488,116,050*	*65,699,950*	
Large-scale studies (50/50)	30,000,000	30,000,000	60,000,000
Total conventionally authorized subtotal	**518,116,050**	**303,799,950**	**821,916,000**
Total LCA Study cost share	**1,256,221,000**	**714,423,650**	**1,995,981,000**
Total cash contributions	*1,283,341,350*	*325,704,400*	
Total Real Estate		386,719,000	

[a]"Conditionally authorized" refers to items that have been "recommended for specific Congressional authorization, with implementation subject to Secretary of the Army review and approval of feasibility-level decision documents" (U.S. Army Corps of Engineers, 2004a).

[b]For the conditionally authorized feature of small diversion at Hope Canal, LERRD exceeded 35 percent of the total project cost by $25,336,250, which is reimbursed to the nonfederal sponsor.

[c]"Programmatically authorized" refers to items Congress has authorized USACE to proceed with the group of projects without authorizing the individual projects involved.

[d]"Conventionally authorized" refers to items proposed in the Chief's Report and authorized by Congress through the Water Resources Development Act.

SOURCE: Modified from U.S. Army Corps of Engineers, 2004a. (Refer to U.S. Army Corps of Engineers [2005b] for cost revisions.)

approximately 86 percent of the total cost is for projects (including the beneficial-use program), and the remainder is dedicated to advancing scientific, engineering, and technical skills. For a state-level cash contribution of more than $325 million, Louisiana would obtain a project worth nearly $2 billion. The total cost associated with compensating landowners alone exceeds the state's cash contribution.

RELATIONSHIP OF COAST 2050 AND THE LCA STUDY TO CWPPRA PROJECTS AND EXPERIENCE

The overall purpose, methodologies, procedures, and participants are generally consistent between CWPPRA and the LCA Study. This is not surprising given that the LCA Study builds on and represents an evolution of the information and lessons learned from CWPPRA.

However, notable differences between the two programs exist. Foremost among the differences is the scale of the projects associated with each of the two programs. With limited funding for a problem that spans the entire Louisiana coast, "CWPPRA has concentrated on small-scale projects distributed across the coast. In contrast, the LCA [Study] focuses on larger projects that would generally work at an ecosystem scale" (U.S. Army Corps of Engineers, 2004a).

Another significant difference between the two programs is the increased emphasis that the LCA Study places on future studies and research, whereas CWPPRA focuses almost exclusively on constructing and monitoring restoration projects. The LCA Study is more comprehensive and forward thinking, as demonstrated by the inclusion of funding for a 10-year S&T Program, demonstration projects, a program to increase the beneficial uses of dredged material, and investigations of modifications of existing structures.

A third significant difference is the extent of public involvement in the programs. CWPPRA has built two formal elements of public involvement into the program: a public outreach committee and citizens participation group. The LCA Study, on the other hand, does not propose any formal structure to ensure public involvement in the future, which is a serious deficiency of the plan.[3]

Project selection is another difference between the programs. CWPPRA and Coast 2050 conducted extensive outreach to the public through parish-level meetings in order to identify needs and potential projects. Input from these meetings was reflected in the restoration plan. Project selection in the LCA Study was based on a series of outputs from computer models and an array of selection criteria, which are examined in detail in Chapter 5. Given the overall extent to which the two programs complement one another and are on such parallel tracks, consideration should be given to integrating the two efforts at an appropriate time.

---

[3]The Louisiana Department of Natural Resources and USACE personnel have been working since March 2005 to develop and implement mechanisms to increase public involvement in any eventual Louisiana coastal area program implementation (G. Duszynski, written communication, 2005).

IMPROVING ONGOING RESTORATION EFFORTS

The history of restoration efforts in Louisiana over the past four decades underscores the magnitude of the challenge facing the state and the nation. Although many of the efforts have led to land gain in limited areas, the overall impact of these efforts has done little to stem the increasing toll of coastal erosion. The recognition that more ambitious approaches would be needed, as envisioned in Coast 2050, represents a significant step in understanding the problem. The draft LCA Comprehensive Study, and the effort that went into its development, represents the type of effort needed to address large-scale land loss in Louisiana. Although the draft LCA Comprehensive Study was not the subject of this review, much of the supporting documentation in the LCA Study was derived to support the draft LCA Comprehensive Study. Thus, although some of the weaknesses of the LCA Study were inherited from the preceding effort, or came about because of the short transition time allowed between the completion of the draft LCA Comprehensive Study and the release of the LCA Study, it has its own shortcomings. Principal among these is the impression that it offers too modest an effort. Unless efforts proposed in the LCA Study successfully lay the groundwork for an ongoing, ambitious effort that extends beyond the next 10 years, that impression may be shown to be valid. Subsequent chapters lay out steps that should be undertaken if the efforts outlined in the LCA Study are to establish requirements for the subsequent and more ambitious activities that will undoubtedly be needed.

5

The LCA Study Planning Approach, Modeling, and Project Selection Process

---

HIGHLIGHTS

This chapter
*   Addresses issues associated with the planning approaches proposed in the *Louisiana Coastal Area (LCA), Louisiana—Ecosystem Restoration Study* (LCA Study) to counter serious land loss rates
*   Examines past and proposed efforts to engage stakeholders
*   Discusses models used in the project selection process and identifies their strengths and weaknesses
*   Examines the overall project selection process

In an area as vast and complex as the Louisiana coastal region, planning for restoration is a challenge. The fields of ecology, wetland science, hydrology, geology, oceanography, computer modeling, engineering, sociology, economics, political science, land-use planning, hazard mitigation, and law can make a contribution in defining the problem and providing possible solutions; therefore, these fields must be considered in the design process.

CONTEXT FOR PLANNING

State Legal System and Issues

Planning takes place in the context of laws governing land-use and property rights, as well as the institutional arrangements of federal, state, and local government. Louisiana, by virtue of its history as a French colony, has its law based on the Napoleonic Code rather than English common law, which is found in the rest of the United States. In the English common law tradition, the judiciary acts as a check on both the executive and legislative branches, limiting their ability to alter contract and property rights. Judges, in this system, use common practice and court precedent to interpret laws. Napoleonic Code takes the civilian law approach, based on scholarly research and the intent of the lawmakers. The French tradition is more comfortable with a centralized and activist government, limiting the judiciary's role to ensuring that the will of the government is enforced (Benjamin, 2001). These differences in philosophy influence judicial decisions regarding compensation of landowners or other aspects of citizen rights.

Most of coastal Louisiana is privately owned or at least subject to some claim of private ownership (Davis, 2002). Much of the coastal land is owned for its underlying minerals (i.e., oil and gas). However, Louisiana claims ownership over navigable water bottoms including lands that have been submerged through erosion or subsidence (U.S. Army Corps of Engineers, 2004a). According to the Louisiana Constitution (Article IX, Section 3), owners of land contiguous to and abutting navigable waters owned by the state "shall have the right to reclaim or recover land lost through erosion, compaction, subsidence, or sea level rise occurring on or after July 1, 1921. Such private efforts to restore or reclaim lost lands can be made at any time" (Act 6, Louisiana Wetlands Conservation and Restoration Act, 1989; U.S. Army Corps of Engineers, 2004a). In the cases of subsided interior marshes, the State of Louisiana does not assert or claim ownership that it could, given the navigable standard.

Coastal restoration projects may impinge on private reclamation rights. The Louisiana Wetlands Conservation and Restoration Act of 1989 provides "that [the Louisiana Department of Natural Resources] may enter into negotiated boundary agreements with such disaffected landowners to address the anticipated loss of their ownership and reclamation rights" (Act 6, Louisiana Wetlands Conservation and Restoration Act, 1989; U.S. Army Corps of Engineers, 2004a) where the LCA Study is anticipating creating land. It is still possible that when land emerges from water bottoms claimed by Louisiana, the previous landowner may attempt to claim that he was deprived of his reclamation rights to the emergent land.

Mineral rights (e.g., oil and gas) are not extinguished when land is flooded. The mineral rights are not acquired when the state or federal government purchases the land by fee simple, which means there are no restrictions on the transfer of ownership. Thus, the mineral rights would be expressly reserved to the previous landowner or to the lessee of those mineral interests. The mineral interest owner or lessee would be allowed to continue ongoing mineral activities. Lands and improvements actually used or destroyed for levees or levee drainage purposes shall be paid for as provided by law (Louisiana Constitution [1974], Art VI, §42).

Louisiana's judicial system has been used to settle claims from landowners suffering damage as a result of coastal restoration activities. A constitutional amendment passed in 2003 limits Louisiana's liability for damages caused by coastal restoration projects to the land's fair market value in line with federal standards, and it applies retroactively (Louisiana Constitution [1974], Amendment I, §42). This amendment was in response to a $2 billion judgment by a state court that sided with the oyster growing and harvesting interests who claimed that their oyster beds were destroyed by the Caernarvon freshwater diversion. The Louisiana Supreme Court reversed the decision. As a result, program feasibility and costs will be contingent upon the legal and legislative rulings and actions that frame the planning exercise. When mineral rights are affected or damages to preexisting shellfish or fishing activities are caused by wetland restoration activities, the planners will have to incorporate institutional changes and financial compensatory elements into the planning process.

Array of Agencies and Interests

Coastal restoration, as envisioned by the LCA Study, involves federal, state, and local governments. The projects being proposed in the LCA Study will impact, both positively and negatively, the finfish and shellfish industry, the gas and oil industry, the petrochemical industry, banking, agriculture, shipping, recreation, and the everyday life of local residents. The projects will require the purchase of land and the resettlement of homes and businesses. Environmental organizations will be involved as habitats are changed and restored. There will be tension between those that have adjusted to the land loss and salinity change by changing their occupations and lifestyle and those that will benefit from land restoration and the increasing freshness of the waterways.

Because dramatic changes are envisioned by the LCA Study, it will alter the way people live, work, and play; those impacted will have to be convinced of the benefits and agree to change their lifestyles. Therefore, the LCA Study is about both technical challenges and social, economic, and political challenges. Consensus building, through education, discussion, and trust, can be partially effective in building the cooperation necessary for coastal restoration to be successful. This cooperation requires outreach by state and federal agencies to educate local governments (elected officials and staff) and to gain the local governments' perspectives. The three levels of government will also have to educate and involve the public, listen to public concerns, and work with the public to overcome inequities.

The Nature of Planning

The planning process utilized in the LCA Study is designed to find the plan that best meets planning objectives. The stated overarching objective is "to reverse the current trend of degradation of the coastal system" (U.S. Army Corps of Engineers, 2004a). The planning process is best described by planning theory as "rational" or "synoptic." The emphasis in this approach is to define the problem, develop strategies to resolve the problem, evaluate alternatives to find the "best" strategy, and then adopt and implement it. This rational planning process is appropriate for complex technical problems such as restoring the coast; however, it is less successful when the problem is "wicked" and highly political. Wicked problems, as described by Rittel and Webber (1973), are ill-defined; often lack consensus regarding their causes, obvious solutions, and criteria for determining when a solution has been achieved; and have numerous, often unknown, interconnections to other processes and problems. To determine whether or not the loss of Louisiana's wetlands is a wicked problem, one has only to ask what the solution to the problem is, and the debate begins regardless of whether the participants are seasoned researchers or laypersons. The fact that the problem is highly political is obvious because whatever the solutions, some people will gain, and others will lose. The political system becomes engaged because in a democracy, politics is used to mediate between those who benefit and those who pay since those who pay the price for improvements may not be the ones that receive the benefits. "Pay" does not just mean money but also means inconvenience, time, and exposure to hazards.

Increasingly, planners have recognized that simply telling people what they have to do because it is the "right" thing does not work well in a democracy. It appears that people have to adopt the ideas as their own and be committed to changing, even if it makes life less convenient and incurs a cost, if major plans and projects are to succeed. Communicative planning, advocated by Innes (1995, 1998), is built on information exchange between experts and citizens. It is not just an exchange of words but reflects a variety of institutional, political, and power relationships. During the course of these exchanges, a shared sense of meaning develops, and subsequent actions will be heavily influenced by this shared understanding (Brooks, 2002). "The planner who follows this approach is not an analyst working behind closed doors to eventually produce the most rational recommendations but an active and intentional participant in a process of public discourse and social change" (Ozawa and Seltzer, 1999). These sociopolitical challenges can be overcome if a robust adaptive management effort, which includes adequate mechanisms for addressing the full range of challenges, is employed.

ROLE OF MODELS IN THE PLANNING AND ADAPTIVE MANAGEMENT OF THE LCA STUDY PLANNING PROCESS

Models have been used in attempts to understand the physical processes within the project area, for the development of the *Louisiana Coastal Area, LA—Ecosystem Restoration: Comprehensive Coastwide Ecosystem Restoration Study* (draft LCA Comprehensive Study), and for project selection and prioritization of actions within the LCA Study, and they are expected to be part of the future Adaptive Environmental Assessment and Management (AEAM) Program called for in the LCA Study. The draft LCA Comprehensive Study was used to create the LCA Study as a near-term alternative to its predecessor. (For additional background, refer to Chapter 3.) These models were assembled in the period August 2002 to September 2003 (U.S. Army Corps of Engineers, 2003a), and model selection was based primarily on those available from previous studies by academia and state and federal agencies and on model integration and development that could be achieved in the available time. If the models are to be an integral part of the management process, it is important that they be credible and robust representations of the environmental and economic processes they are intended to represent.

In complex long-term projects, modeling[1] can be a valuable aid to managers for the following reasons:

• Generate agreement on the important processes determining the response of a system to natural and anthropogenic perturbations

---

[1]General characterization of a process, object, or concept in verbal or mathematical terms, which enables the relatively simple manipulation of variables to be accomplished in order to determine how the process, object, or concept would behave in different situations (Office of Administrative Law Judges Law Library, 1991).

- Project future conditions under different scenarios, including "no action"
- Allow performance criteria to be quantified and tracked
- Develop temporal and spatially distributed understanding not possible by monitoring alone
- Optimize data collection programs
- Evaluate sensitivity and uncertainty
- Clarify the assumptions and their influence on results
- Facilitate cross-disciplinary communication and explain processes to nontechnical audiences
- Assist real-time emergency management during severe flood or storm surge conditions

The use of models and a sensor network for real-time emergency response is self-evident and should be addressed by the science coordination team (U.S. Army Corps of Engineers, 2004a) as the suite of models for the LCA Study are refined in the future; thus, they are not explored further in this report. The central role of modeling in the proposed AEAM Program (Figure 5.1) is described in the LCA Study and discussed further in this report. The use of models in predicting whether the future preferred landscape dynamic is achievable and how projects should be selected is discussed in Chapter 6. The suite of models used to develop the linkages between physical processes and ecological response is shown in Figure 5.2.

Types of Models

The LCA Study modeling team differentiates between two types of modeling efforts (Figure 5.3). The first is simulation modeling, which is predictive, deterministic, and process-based (U.S. Army Corps of Engineers, 2003a). "Simulation modeling represents the highest level of sophistication in ecological modeling where clearly defined assumptions of ecological mechanisms are linked to geophysical process. These models can be used to simulate the endpoints of engineering alternatives" (U.S. Army Corps of Engineers, 2003a). The hydrological models are simulation models. Less sophisticated are the hydrological box models that can predict endpoints of salinity, hydroperiod, and sedimentation over longer time scales and at a coarse spatial resolution.

In the second type of modeling used in the LCA Study process, empirical information was used to statistically estimate ecosystem responses to various changes in the environment. This type of model is called a desktop model. The coarse-scale "desktop" statistical approach might use spreadsheets and is based on reported relationships in various literature

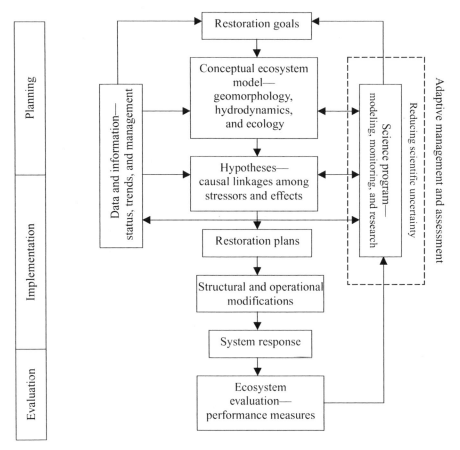

FIGURE 5.1 The Science and Technology Program approach proposed for developing comprehensive ecosystem restoration plans for the LCA Study (U.S. Army Corps of Engineers, 2004a; used with permission from the U.S. Army Corps of Engineers).

sources, data, and best professional judgment. A goal of the LCA Study is to eventually develop and use simulation models for all five modules across all four regions (U.S. Army Corps of Engineers, 2003a).

Goal of Modeling

The goal of the modeling effort was to develop quantifiable benefits, or outputs, of the plan based on purely ecological criteria. The following are these criteria:

- Land building measured in acres
- Habitat switching measured as change of habitat types in acres
- Primary productivity of land and water measured as an index of composite plant productivity
  - Habitat use measured in habitat units for selected species
- Removal of nitrogen from the Mississippi River measured as a percentage of nutrients removed (U.S. Army Corps of Engineers, 2004a)

Specifics of the Models

The LCA Study's ecosystem model was the product of a team of more than 38 scientists, engineers, and resource managers, primarily from Louisiana. Subgroups of technical experts were assembled in a workshop to develop algorithms for each of the five modules of the ecosystem model (Figure 5.2). Because of the size of the study area and the schedule, the model was a hybrid of simulation and desktop modeling (Figure 5.3). These subgroups were responsible for integrating expertise among university and agency scientists and interfaced closely with the LCA Study managers to develop appropriate modeling scenarios.

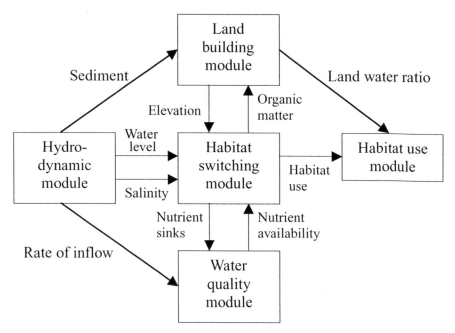

FIGURE 5.2 Linkage of different modules used in desktop and simulation models (U.S. Army Corps of Engineers, 2004a; used with permission from the U.S. Army Corps of Engineers).

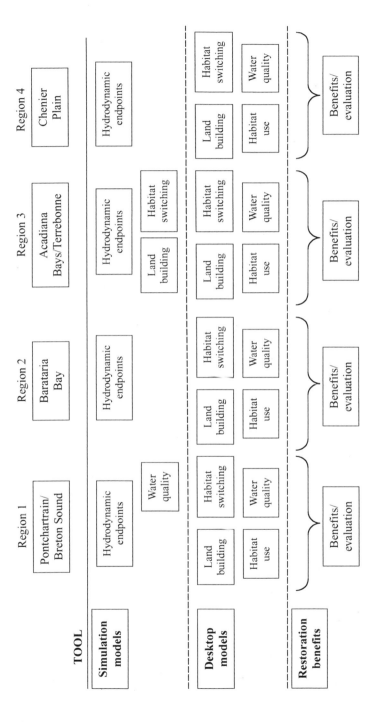

FIGURE 5.3 Hybrid of desktop and simulation modeling tools for benefit evaluation used in the LCA Study (U.S. Army Corps of Engineers, 2004a; used with permission from the U.S. Army Corps of Engineers).

A spatial orientation was included in the modeling of ecosystem responses. The study area was divided into 1 square kilometer (km^2) (0.39 square miles [mi^2]) polygons resulting in 43,138 coastal area cells. Although 1 km^2 (0.39 mi^2) is reasonable for modeling, it results in problems of land classification and resolution of processes at subgrid scale. For example, a cell was identified as a "water cell" if it contained more than 0.40 km^2 (0.15 mi^2) of open water, with all other cells identified as non-water and classified as the vegetation type that covered the most area in the cell (U.S. Army Corps of Engineers, 2003a).

Additionally, each of the four LCA Study's regions was partitioned into boxes. Either boxes were constructed to fit previous conventions of box construction, or vegetation maps were used to distinguish zones of salinity regimes in the coastal landscape. A major assumption is that all the land cells of each box will respond to the results of nodes in respective boxes.

The LCA Study's ecosystem model is made up of a number of different models (or modules) that feed into one another (Figures 5.2 and 5.3). This generalized picture does not reflect the detail of the modules. For example, the hydrodynamic portion of the ecosystem model comprises five different models selected from previous studies (Figure 5.3):

• Region 1—Princeton Ocean Model was developed by Blumberg and Mellor (1987). This is a three-dimensional sigma coordinate primitive variable model. Data were available for validation only for the Lake Pontchartrain area (Figure 5.4).

• Region 2—TABS-MD (RMA 2, RMA 4) developed by the U.S. Army Corps of Engineers (USACE). This model relies on a central two-dimensional, finite-element representation of estuarine and fluvial hydrodynamics and was used for the lower two-thirds of the Barataria Basin.

• Region 3—Coastal Ecological Landscape Spatial Simulation model (Sklar et al., 1985; Costanza et al., 1987, 1989) was the initial model applied to Terrebonne watershed. This model evolves the landscape composition according to explicit rules and formulas, and results are calibrated against the record of past landscape change and forced by a continuous hydrodynamic simulation (Martin et al., 2000; Reyes et al., 2003).

• Region 4—H3D (Stronach et al., 1993) is a three-dimensional hydrodynamic and advection dispersion model used to study the Calcasieu-Sabine Basin. MIKE-11, produced by the DHI (Danish Hydraulic Institute) Water and Environment, was used to simulate flows, sediment transport, and water quality in the Rockefeller Refuge. It is composed of two dynamically linked modules—the hydrodynamic module and the advection dispersion module.

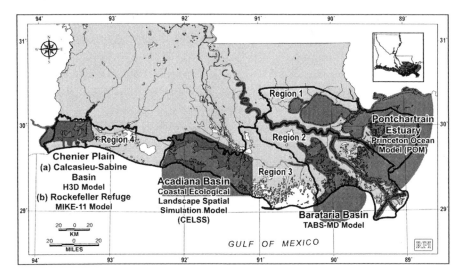

FIGURE 5.4 Deterministic models of the Louisiana coastal area (background map supplied by Research Planning, Inc.).

The objectives of the original model studies varied for the regions and covered differing amounts of the coast. Figure 5.4 illustrates that large gaps within the Louisiana coastal area are not covered by deterministic models. Because of the differing objectives of the models selected, there is no consistency in the selection of bathymetry (where available), period of time the models simulate, and boundary conditions. However, this preliminary phase of modeling has identified gaps in essential data, demonstrated the most important processes that should be included in a model formulation, and generated experience in the use of a broad range of one-dimensional, two-dimensional, and three-dimensional models. These experiences will be invaluable when the science coordination team (Figure 4.5) establishes a consistent modeling approach for the entire project area.

Table 5.1 lists all of the variables that were used in the five general models: hydrodynamics, land change, water quality, habitat switching, and habitat use.

The compilation of the conceptual approach and the model application represent a significant amount of modeling work and coordination to quantify the outputs of each of the features and frameworks. The integrated modeling exercise has been ambitious, and the modeling team has achieved much and should be commended for articulating a comprehen-

TABLE 5.1 Module Variables and Interactions Among Modules

Variable	Module				
	Hydrodynamics	Land Change	Water Quality	Habitat Switching	Habitat Use
Wind speed and direction	Input				
Initial water level	Input				
Initial salinity	Input	Input[a]			
Initial temperature	Input				
River temperature	Input				
Historical land change rates		Input			
River sediment load		Input			
Sediment retention factor		Input			
Bulk density of deltaic soils		Input			
Initial land area	Input	Input			
Bathymetry	Input	Input	Input		
Land elevation	Input	Input	Input	Input	Input
Diversion flows	Input	Input	Input		
River nitrogen			Input		
Nourishment factor	Output[a]	Input			
Salinity	Output		Input	Input	Input
Water level	Output		Input	Input	Input
Water residence time	Output		Input		
Water temperature	Output		Input		Input

Wetland area	Output			Input
Habitat type			Input	Input
Nitrogen removal		Output		
Water primary production		Output		
Wetland primary production			Output	
Chlorophyll a water column[b]				
Habitat quality, alligator				Output
Habitat quality, dabbling duck				Output
Habitat quality, mink				Output
Habitat quality, muskrat				Output
Habitat quality, otter				Output
Habitat quality, Atlantic croaker				Output
Habitat quality, brown shrimp				Output
Habitat quality, gulf menhaden				Output
Habitat quality, largemouth bass				Output
Habitat quality, oyster				Output
Habitat quality, spotted seatrout				Output
Habitat quality, white shrimp				Output

[a] In regions 1, 2, and 3, the nourishment factor refers to freshwater and is based on the initial salinity of the receiving basin. In region 4, the nourishment factor is based on the change in salinity relative to the no-action scenario.

[b] No value provided in source.

SOURCE: U.S. Army Corps of Engineers, 2003a.

sive approach. The modeling team[2] has clearly described and referenced the models in an attempt to be as transparent as possible. Uncertainties and gaps in knowledge have been systematically identified (U.S. Army Corps of Engineers, 2003a). In the very short time frame (13 months) of this initiative, much has been achieved in assembling data and understanding how results will be interpreted.

Challenges

Several challenges are posed by the current approach to modeling that distinguish this effort from similar studies elsewhere in the United States. These challenges are as follows:

- The study area, even when broken into the four regions, is still at a massive spatial scale for simulation models. Keeping track of 43,138 units, each with an area of 1 km^2 (0.39 mi^2), can be overwhelming.
- The landscape is difficult to work in due to problems (private lands and rough terrain) in accessing all sectors. This makes it difficult to accurately characterize open boundaries, and small errors can be magnified as they cascade through the linked models.
- The problem of coastal land loss is severe and immediate so restoration decisions have been, and will continue to be, made before model calibration and algorithm development can occur. For example, no sensitivity analyses have been run on the land-building module, habitat-switching module, water-quality module, or habitat-use module that are being used to predict outputs for different projects in coastal restoration. Decisions are being made based on these outputs, which could be largely erroneous.
- There are limited data in many key areas. The modeling team documents a number of key data gaps (Appendix C of U.S. Army Corps of Engineers [2004a]), including no survey data for the four major lakes (White, Grand, Calcasieu, and Sabine), inadequate marsh elevation data throughout the region, and extremely limited hydrologic and ecologic data for the Region 4 marsh.
- Several fundamental questions are yet to be resolved; however, the Science and Technology (S&T) and AEAM programs are proposed to deal with them. Some of these questions involve the transport of sediment over long distances using pipes, the effect of using marine sands on

---

[2]The modeling team was part of the working groups (Figure 4.4) and will be part of the Science and Technology Program (see Chapter 6) and the AEAM framework (see Chapter 6).

marshes, and the amounts and kind of sediment carried by the Mississippi River in different parts of its water column and how to use that variation to the best advantage.

Uncertainties

The draft LCA Comprehensive Study (U.S. Army Corps of Engineers, 2003a) systematically identified uncertainties[3] and gaps in knowledge. Uncertainties are present in different modules in terms of the model rigor and the quality of the parameter estimates in model input. Many of the models used professional judgment as the data source (U.S. Army Corps of Engineers, 2003a). Model developers acknowledge that many of the data are of low quality and few of high quality, and they report that model rigor is moderate to low (U.S. Army Corps of Engineers, 2003a). Given the way the models were used and the multiple levels of information exchange and interaction, the errors and uncertainties will propagate through this sequence of models. There is a need to estimate the combined effects of these uncertainties on the projected outcomes and thus on the derived benefits. Appropriate project selection depends on the validity of these models, and it does not appear that the modeling results discussed in the draft LCA Comprehensive Study were modified in any way before being used for input to project selection in the LCA Study.

PROJECT SELECTION AND THE LINK WITH MODELING

The draft LCA Comprehensive Study and the LCA Study both outline in detail the methodology for project selection (U.S. Army Corps of Engineers, 2004a) (Figure 5.5). The project delivery team, which consisted of state and federal agency representatives, developed core restoration strategies for each region. Restoration "features," specific projects or a

---

[3]The term "uncertainty" has a variety of definitions and uses. When dealing with numerical modeling of natural systems or economic analysis, it often refers to statistical uncertainty. For example, risk analysis from flooding uses probabilistic descriptions of the uncertainty in estimates of important variables, including flood-frequency, stage-discharge, and stage-damage relationships, to compute probability distributions of potential flood damages. These computed estimates can be used to determine a levee height that provides a specified probability of containing a given flood. The LCA Study often uses the term "uncertainty" in the broader sense and interchangeably with the term "knowledge gap" to describe instances in which insufficient understanding of underlying processes may exist. Here and throughout this report, the term "uncertainty" is used to refer to statistical uncertainty associated with a parameter of interest. The term "knowledge gap" is used to refer to an area in which understanding may be insufficient to support effective decision making.

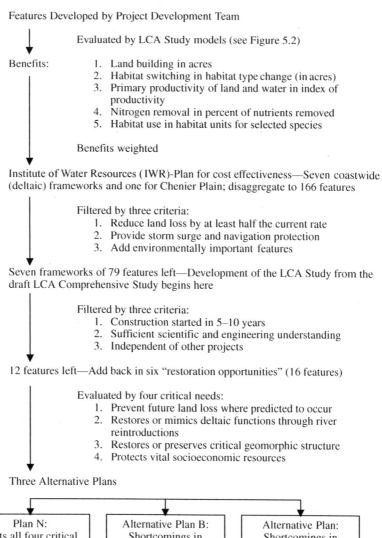

Features Developed by Project Development Team

Evaluated by LCA Study models (see Figure 5.2)

Benefits:
1. Land building in acres
2. Habitat switching in habitat type change (in acres)
3. Primary productivity of land and water in index of productivity
4. Nitrogen removal in percent of nutrients removed
5. Habitat use in habitat units for selected species

Benefits weighted

Institute of Water Resources (IWR)-Plan for cost effectiveness—Seven coastwide (deltaic) frameworks and one for Chenier Plain; disaggregate to 166 features

Filtered by three criteria:
1. Reduce land loss by at least half the current rate
2. Provide storm surge and navigation protection
3. Add environmentally important features

Seven frameworks of 79 features left—Development of the LCA Study from the draft LCA Comprehensive Study begins here

Filtered by three criteria:
1. Construction started in 5–10 years
2. Sufficient scientific and engineering understanding
3. Independent of other projects

12 features left—Add back in six "restoration opportunities" (16 features)

Evaluated by four critical needs:
1. Prevent future land loss where predicted to occur
2. Restores or mimics deltaic functions through river reintroductions
3. Restores or preserves critical geomorphic structure
4. Protects vital socioeconomic resources

Three Alternative Plans

| Plan N: Meets all four critical and needs plan that best meets the objectives | Alternative Plan B: Shortcomings in restoring critical geomorphic structure | Alternative Plan: Shortcomings in addressing Mississippi River reintroductions |

Array features and opportunities on how fast can start
Apply $1.9 million

Get five projects

FIGURE 5.5 Overview of the project selection process (U.S. Army Corps of Engineers, 2004a; used with permission from the U.S. Army Corps of Engineers).

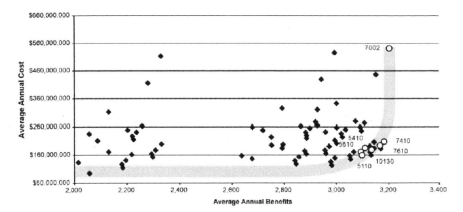

FIGURE 5.6 Preliminary average annual costs and average annual benefits for the final array of alternative frameworks for Regions 1, 2, and 3 (U.S. Army Corps of Engineers, 2004a; used with permission from the U.S. Army Corps of Engineers). (NOTE: The gray line denotes the cost efficient frontier; projects that fall at the bend in the frontier provide the maximum benefit for the least cost.)

collection of elements, were then considered in more detail; a total of 166 features were considered across four regions. Features were selected based on Coastal Wetlands Planning, Protection, and Restoration Act (CWPPRA) experiences and professional judgment. These features were combined into "frameworks" or plans.[4] Hydrodynamic and ecological models were then used to assess the ecological benefits of each framework. The outcomes of these models provided input for the economic model Institute of Water Resources (IWR) Plan Decision Support Software, which was used in the draft LCA Comprehensive Study and the LCA Study to establish frameworks with the lowest cost for any benefit level (Figure 5.6). Those frameworks that emerged on the cost-efficient frontier of the cost versus outputs measure were then disaggregated into features, and the features were screened by a series of criteria. The final features were ordered based on the time required for implementation, and those consistent with the available funding became the top five projects. These steps are reviewed carefully in this section because this process forms the basis for project selection for the LCA Study and is assumed to be the prototype for future plans.

---

[4]Features could appear in several frameworks such that each framework was simply a unique collection of features.

As shown in Figure 5.5, the models just discussed were used to calculate benefits that would be developed as a result of undertaking a "framework" or plan. The benefits were all ecological in focus (e.g., acres of habitat, percentage of nitrogen removed).

It would not have been too difficult, although perhaps more arbitrary, to include a benefits dimension for socioeconomic value as a complement to the habitat-use measure, where the "species" is humans. For example, an index of the proximity of a project area to a major urban settlement, an index of navigation protection, or the reduction in hurricane risk or river flooding would be measures that could be incorporated for cost-effectiveness analysis. Inclusion of these types of indices would probably skew project selection toward projects having more beneficial socioeconomic impacts, compared to using the initial ecological selection criteria with a "socioeconomic critical need" criterion applied only in the final steps of the analysis. Such an analysis, however, is not an essential element of projects proposed as National Environmental Restoration (NER) but is more consistent with the analysis employed when evaluating projects proposed as National Economic Development (NED).

A framework's effectiveness was measured by a weighted sum of the five benefits (presented earlier in the "Goal of Modeling" section). The principle of relative weightings is valid and should reflect the importance of a benefits category to overall program goals. Appropriate weighting of selected outcomes used in the study, however, is not evident. For instance, does the habitat-switching measure represent a positive output (e.g., is the conversion of 1 acre [0.0040 km^2; 0.0016 mi^2] of salt marsh to fresh marsh given a +1 or a −1)? A human-based weighting may differ substantially from a purely ecological-based weighting. Removal of nitrogen may be highly important to the fisheries but not to storm protection. Does the weighting imply something about the importance of various socioeconomic activities dependent on the coast? Furthermore, it is not evident how the ecological-based weights were selected, because weights should reflect the value of an output with respect to a goal, and the purely ecological goal of an ecosystem is not evident. In any case, given these reservations, it would be mandatory to test different weighting schemes in order to determine how sensitive the prioritization of frameworks and their projects is to these weightings. No evidence is presented in the LCA Study that this was done.

A more technical, but important, consideration in the benefits analysis process is whether the scoring of the different benefit dimensions was scaled appropriately so that benefit categories would not receive undue weightings. For example, was each of the scoring dimensions scaled as 0–1 or 0–100? This scaling is important; otherwise, a combination of misscaled measures plus the importance weighting would result in total benefits scores of projects that reflect a meaningless combination of relatively arbitrary scoring and importance. The technical studies describing the ecological or physical scoring did not describe the scaling applications; thus, it is not clear how the scaling issue was addressed.

Economic Analysis

Using project costs and these weighted benefits (output) measures by framework, USACE used the computer model IWR Plan to establish combinations of frameworks and their cost-effectiveness. These costs and benefits are shown in Figure 5.6.

The cost-efficient frontier defines those combinations of frameworks with the lowest cost for any benefit level and the highest benefit for any cost level. The LCA Study consisted of frameworks selected from, or near, this frontier. Several of these cost-effective frameworks were excluded for varying reasons. For example, some could be funded under CWPPRA, and others would not provide substantial storm surge protection. These exclusions resulted in six plans, and two more were added for a total of eight.

Filtering Criteria

Project framework combinations that met the cost-effectiveness were next evaluated using the following three criteria:

1.   Would implementation reduce land loss by at least half the current rate?
2.   Would implementation provide storm surge protection and protect navigation?
3.   Would implementation add environmentally important features?

One framework was eliminated at this point (no explanation provided), leaving seven combinations for further consideration (U.S. Army Corps of Engineers, 2003a). These seven frameworks consist of 79 features out of the original 166. Socioeconomic values of storm and navigation protection are introduced as sorting criteria at this stage. This is also where projects for inclusion in the draft LCA Comprehensive Study stopped. Therefore, the following discussion applies to how information developed for the draft LCA Comprehensive Study was used to formulate the near-term priorities laid out in the LCA Study.

The 79 projects were then sorted (filtered) for their inclusion in the LCA Study, first using the following criteria:

1.   Engineering and design complete and construction started within 5–10 years (reduced 79 eligible features to 61 features)

2.   Sufficient scientific and engineering understanding of processes (reduced 61 eligible features to 33 features)

3.   Implementation is independent and does not require that another restoration feature be implemented first (reduced 33 eligible features to 12 features)

Several features not meeting the independence test were combined to represent interdependent packages, adding six restoration opportunities (consisting of 16 features) to the remaining 12 features for a total of 18 projects that were evaluated further (U.S. Army Corps of Engineers, 2004a). The 12 features and six restoration opportunities passing the sorting test were then evaluated by the "critical needs" criteria, which are as follows:

1.   Would implementation prevent future land loss where it is predicted to occur?

2.   Would implementation restore (or mimic) fundamentally impaired deltaic functions through river water and sediment reintroductions?

3.   Would implementation restore or preserve endangered critical geomorphic structures?

4.   Would implementation protect vital socioeconomic resources (including communities, infrastructure, business and industry, and flood protection)?

The frameworks and their component features that met one or more critical need were selected for further analysis (U.S. Army Corps of Engineers, 2004a). For example, the Mississippi River Gulf Outlet (MRGO) restoration feature met critical needs Criteria 1, 3, and 4. Seven features and five restoration opportunities (made up of 14 restoration features) met the critical needs criteria (U.S. Army Corps of Engineers, 2004a).

It is not clear how a feature was determined to meet the criteria to "protect vital socioeconomic resources" (U.S. Army Corps of Engineers, 2004a). For example, the MRGO restoration feature is characterized as protecting "developments located adjacent to MRGO" (U.S. Army Corps of Engineers, 2004a). Although at the time of this writing, it was not clear whether MRGO played a role in flooding St. Barnard Parish during Katrina, prior to Katrina various parties had speculated that MRGO might act as a major conduit for coastal storm surges that could inundate New Orleans. Though never intended to be an effort to reduce storm damage, it is not clear what protection the MRGO restoration feature would have offered during Katrina because the proposed efforts would have taken place south and east of the levee failure responsible for the flooding of St. Bernard Parish. The Maurepas Swamp feature is characterized as protecting "the growing ecotourism industry" (U.S. Army Corps of Engineers, 2004a). How and why these projects provide vital socioeconomic characteristics are not evident. As an example, the decision to protect ecotourism rather than oil and gas infrastructure or communities in other regions is not clear. The critical needs criteria are acceptable, in principle, but the apparently ad hoc manner in which they were applied is ripe for prejudicial and political decision making and is vulnerable to future challenges.

This lengthy cost-effectiveness and sorting criteria methodology resulted in the selection of the projects termed the "plan that best meets the objectives" (U.S. Army Corps of Engineers, 2004a). This plan consists of 24 features, some of which are combined to yield 14 projects. These 14 projects are then ordered based on the time required for implementation. The top five features were selected based on funding availability. The final five projects, discussed more fully in Chapter 6, were recommended for conditional or programmatic authorization and were estimated to cost a total of $864 million (U.S. Army Corps of Engineers, 2004a). These projects are as follows:

1.   MRGO environmental restoration features ($108.3 million,[5] 12.5 percent of the total cost)

2.   Small diversion at Hope Canal ($70.5 million,[6] 8.2 percent of the total cost)

3.   Barataria Basin barrier shoreline restoration ($247.2 million,[7] 28.6 percent of the total cost)

4.   Small Bayou Lafourche reintroduction ($144.1 million,[8] 16.7 percent of the total cost)

5.   Medium diversion with dedicated dredging at Myrtle Grove ($294 million,[9] 34.0 percent of the total cost)

---

[5]USACE, in the 2005 Chief's Report, updated the cost of the proposed MRGO feature to be $105.3 million (U.S. Army Corps of Engineers, 2005b).

[6]USACE, in the 2005 Chief's Report, updated the cost of the small diversion at Hope Canal to be $68.6 million (U.S. Army Corps of Engineers, 2005b).

[7]USACE, in the 2005 Chief's Report, updated the cost of the Barataria Basin shoreline restoration feature to be $242.6 million (U.S. Army Corps of Engineers, 2005b).

[8]USACE, in the 2005 Chief's Report, updated the cost of the small Bayou Lafourche reintroduction to be $133.5 million (U.S. Army Corps of Engineers, 2005b).

[9]USACE, in the 2005 Chief's Report, updated the cost of the medium diversion at Myrtle Grove to be $278.3 million (U.S. Army Corps of Engineers, 2005b).

Although these projects have been selected as a result of making it through a complicated set of selection processes, it remains disconcerting that the socioeconomic trade-offs inherent in rational project selection were not incorporated earlier and more formally in the selection process. If socioeconomic needs had been incorporated at the cost-effectiveness stage, it is likely that the final set of projects would have been quite different.

Furthermore, the cost-effectiveness analysis was carried out on groups of restoration features referred to as frameworks (individual features may require one or more projects to construct, while groups of features comprise restoration "frameworks"). The process began with 166 individual projects (features or measures). The project delivery team combined them into 5,670 alternatives[10] based on combinations of individual projects.

The alternatives were then tested (using the IWR software) for their combinability. Out of a total of 5,670 possible combinations of alternatives, only 139 were deemed combinable. Then the 139 were tested for cost-effectiveness, and 14 emerged as nearly cost-effective; a final seven were then selected for further analysis. These seven consisted of 79 of the initial 166 projects. The 79 remaining projects were then evaluated and possibly eliminated based on considerations such as whether the engineering and design work would allow construction to begin within 5–10 years. This resulted in a final 12 projects, but some "combined features" were added for a total of 18. These 18 were then ranked based on the speed of implementation. The top five were selected based on available funding.

The project selection process primarily uses ecological benefits early on in project formulation and then uses least cost-alternatives for restoration frameworks as a filtering criteria to accept and reject frameworks and projects based upon their socioeconomic value. However, since the physical and ecological relationships between projects in a framework are not clear and frameworks optimized for cost include many projects that are not chosen for implementation, the actual role of socioeconomic factors in project selection is not clear.

Furthermore, although the cost-effectiveness analysis was carried out on frameworks, the selection decision was made for individual features. Since the cost-effectiveness was calculated for groups, there appears to be some potential for individual features that might score poorly if singled out during a cost-effectiveness analysis to be elevated by more cost-effec-

---

[10]An alternative is a combination of features.

tive projects in the same group. Since the selection process then breaks the groups down into individual features, it would seem more appropriate to consider the cost-effectiveness of individual features, unless the projects can be shown to be physically, ecologically, or logistically interrelated. The rational for this analysis is poorly articulated in the LCA Study, reinforcing the need for greater transparency. Obviously, if the more comprehensive approach called for in this report were used, determining the cost-effectiveness of a single project in the absence of all others would not be appropriate.

Increasing the scale of projects that have clear socioeconomic benefits, such as the protection of New Orleans, apparently was not considered. Although large compared to most CWPPRA projects, in fact, most of the features appear to be "small" projects. Since ecological and economic success might be greater or more certain with larger projects, why was the focus not on a few larger projects? Would a big initiative to save and restore barrier islands coupled with several large strategically located diversions not be more effective and informative than many "small" projects?

Planning Within a River Basin or Coastal System Context

As discussed extensively in the National Research Council's report *River Basin and Coastal System Planning within the U.S. Army Corps of Engineers*:

> ... water resources planning and management in a river basin and coastal system context require an integrated approach to provide a balanced consideration of objectives and potential impacts at relevant time and space scales. The need for such an approach is widely endorsed by the water resources planning and management community (National Research Council, 1999a,b). [USACE] has embraced these principles and, in many cases, has played a major role in their definition and in the development of supporting methods (U.S. Army Corps of Engineers, 1999b). (National Research Council, 2004a)

A 1999 policy guidance letter defines USACE policy as a watershed approach:

> [USACE] will integrate the watershed perspective into opportunities within, and among, civil works elements. Opportunities should be explored and identified where joint watershed resource management efforts can be pursued to improve the efficiency and effectiveness of the civil works programs. [USACE] will solicit participation from federal, tribal, state, and local agencies, organizations, and the local community to ensure that their interests are considered in the formulation and implementation of the effort. Due to the complexity and interrelation of sys-

tems within a watershed, an array of technical experts, stakeholders, and decision makers should be involved in the process. This involvement will provide a better understanding of the consequences of actions and activities and provide a mechanism for sound decision making when addressing the watershed resource needs, opportunities, conflicts, and trade-offs. (U.S. Army Corps of Engineers, 1999b)

As described by the National Research Council (2004a):

In this context, the term "watershed" is interpreted by [USACE] and others to indicate not only terrestrial watersheds, but also coastal systems. This connection is made explicit in other [USACE] guidance and policy documents, such as those describing the Regional Sediment Management Program (e.g., Martin, 2002; U.S. Army Corps of Engineers, 2002a,b). [USACE's] commitment to a watershed approach is also formalized in the *Unified Federal Policy for a Watershed Approach to Federal Land and Resource Management* (65 Fed. Reg. 62566, October 18, 2000), which was adopted by USACE and other federal agencies. Clear support for an integrated planning approach has also been provided by [USACE] leadership. (National Research Council, 2004a)

In testimony before the U.S. Senate, Chief of Engineers Lieutenant General Robert Flowers stated:

Right now, existing laws and policies drive us to single focus, geographically limited projects where we have sponsors sharing in the cost of the study. The current approach narrows our ability to look comprehensively and sets up inter-basin disputes. It also leads to projects that solve one problem but may inadvertently create others. Frequently, we are choosing the economic solution over the environmental when we can actually have both. I believe the future is to look at watersheds first; then design projects consistent with the more comprehensive approach. (U.S. Senate, 2002)

The framework analysis that constitutes much of the project selection in the draft LCA Comprehensive Study and the LCA Study reflects an attempt to identify groups of projects that may represent least-cost alternatives for coastal restoration. In general, the emphasis placed on planning within a watershed or coastal system context is derived from concerns that interactions among the natural processes altered by water resource projects may have unforeseen and adverse consequences. It is also possible to design projects so that interactions among altered natural processes may have synergistic, beneficial effects. As discussed in greater detail in Chapter 6, the five restoration features selected and proposed in the LCA Study do not appear to reflect an attempt to optimize project selection to maximize positive synergistic effects.

THE IMPROVED MODELING AND
PROJECT SELECTION PROCESS

The modelers undertook a challenging assignment to link distributed physical process models and desktop models to develop methods for determining the benefits of a diverse range of actions over a very large area. The individuals involved in the effort are to be commended for articulating the limitations of the modeling effort, as well as acknowledging the need for an S&T Program that will work to fill the knowledge gaps and quantify the performance of the LCA Study.

The process of prioritizing large groups of projects for the LCA Study is generally based on the analysis of individual small projects. With the exception of the Barataria Basin barrier shoreline restoration project, the projects are small in scope, and the end result reflects the fact that only small projects were put into the analysis process, resulting in recommendations of small projects. Serious questions are raised about the narrow vision of this project selection process to identify near-term projects when the overall philosophy has been stated as one of holistic, integrated, problem solving.

The project selection process is based primarily on measures of ecological change, such as land created, nutrient flows decreased, and species protected. This is ecologically-based management. However, some projects will have more economic value than others. Restoring wetlands that function to protect major urban areas would have more economic value than restoring wetlands that provide no such protection. The socioeconomic significance of any project is important to the rational use of resources and to its political acceptability. This dimension should be built into the project selection process earlier rather than left as a criterion filter add-on at the end. Scoring of projects based on socioeconomic significance can be done and would likely be as reliable as scoring based on expected land created or habitat units created for selected nonhuman species. This scoring would provide a rational economic input to project selection.

The draft LCA Comprehensive Study and the LCA Study make repeated references to the importance and economic value of the fisheries dependent on Louisiana wetlands. Yet, the LCA Study has not shown how the five proposed projects will, or will not, affect these valuable fisheries. Brown shrimp and white shrimp, for example, are affected differently by the shifts in salinity caused by freshwater diversions into the wetlands. However, since oysters require higher salinity than persists with freshwater diversions, the oyster culturing industry is adversely affected by some of the restoration efforts. To the extent that protection of the economic vitality of the coastal communities is an objective of the restoration planning effort, the biological and economic consequences need to be stated explicitly.

The basic issues of coastal degradation are "sediment, sand, and salt"—too much of some, too little of others, and in the wrong places. Very broadly, the ecosystem needs protection from Gulf processes below the system and increased inputs of sediments and freshwater from above. The ecosystem needs these protections and inputs at massive ecological scales. Trying to address problems of this scale with a series of small projects is likely to make little difference relative to the scales at which ecosystem changes must occur. It is appropriate to select the small projects as a basis for learning how the system works and for helping to build the confidence necessary for large project selection. However, these small projects produce only small wetland restoration benefits. Is so little truly known about this ecosystem that it is imprudent to implement the large-scale projects that will likely be necessary to save the coast? Or are the small projects selected in order to navigate through the political obstacles that might derail efforts if focus is shifted to larger, more significant projects?

To be useful, models not only must be technically robust but must instill public confidence. Thus, it is important that the models used are defensible, accessible, and transparent so they can be used with confidence. **The model codes employed should reflect widely accepted and verified approaches with a community-wide effort at model development and maintenance. The models should also utilize open-source codes with an active program of model refinement that includes quality control, consistent data sets by all users, and appropriately available and useful data. The management of data, tracking of model data sets, calibration of model parameters, and interaction and coordination of model users and developers are important aspects that should be included in the management plan.** This effort should be structured to attract synergistic collaborations among modelers worldwide and to enhance the current extensive regional expertise in federal and state agencies and academia.

The project selection process primarily used ecological benefits early on in project formulation and then used least-cost alternatives for aggregates of projects (i.e., frameworks) as filtering criteria to accept and reject frameworks based on their socioeconomic value. However, since the physical and ecological relationships between projects in a framework are not clear and the frameworks selected based on cost included many projects that are not chosen for implementation, the actual role of socioeconomic factors in project selection is not clear.

For example, these criteria and the need to demonstrate solid near-term success likely resulted in the avoidance of bold innovative projects that (1) effect a significant sediment delivery to the system, such as abandonment of the Birdsfoot Delta; (2) maximize synergistic effects for reducing land loss over longer time scales by the selection of strategically located or larger-scale projects; or (3) address some of the difficult issues associated with stakeholder response. While the efforts preceding the LCA Study achieved a laudable degree of unanimity among stakeholders on the conceptual restoration plan, this unanimity will be tested by the difficult decisions associated with implementation of the larger-scale projects necessary to achieve greater sediment, water, and nutrient delivery more effectively over a larger area. **The project selection procedure requires more explicit accounting of the synergistic effects of various projects and improved transparency of project selection to sustain stakeholder support. Furthermore, beneficial, synergistic interaction among projects cannot be assumed but should be demonstrated through preconstruction analysis.**

6

The LCA Study and the Feasibility of Its Components

HIGHLIGHTS

This chapter
- Describes and reviews various *Louisiana Coastal Area (LCA), Louisiana—Ecosystem Restoration Study* (LCA Study) elements, including economic analysis
- Provides an overview of the LCA Study's feasibility
- Discusses the modeling strengths and uncertainties and recommends a protocol for future conduct of the modeling program
- Addresses the role of adaptive management and examines management alternatives
- Suggests Third Delta alternatives, including a large, more southern diversion and a full abandonment of the Birdsfoot Delta

The activities proposed in the LCA Study are intended to be a series of interactive efforts that develop the necessary knowledge base to inform longer-term and possibly more expansive restoration efforts, while making some near-term progress by undertaking relatively low-risk projects that will result in a tangible development of what the LCA Study refers to as near-term critical restoration features. The success of these parallel efforts to slow land loss, while developing information needed to support more aggressive project design, is a key component of the strategic approach championed in the LCA Study. As can be seen in Table 6.1, significant resources have been requested to further the study of various scientific and engineering challenges and to help integrate the results and other lessons learned into the overall program management process. The commitment to continued study further underscores the emphasis the LCA Study places on developing a two-pronged approach.

The principal restoration efforts proposed in the LCA Study are directed at the development of five restoration features, where each feature may be completed through either a single project or a series of individual projects. The conditionally authorized cost of the five features, including land easements, rights of way, relocation, and disposal, is estimated to be roughly $864 million. Development of the underlining knowledge base to undertake more robust restoration efforts comprises three additional components of the LCA Study, with a combined budget of $300 million. The first two components include applied research (carried out as part of the Science and Technology [S&T] Program with an estimated budget of $100 million) and formally described demonstration projects (with an estimated budget of $100 million). Lessons learned in dealing with emerging challenges and information developed through the demonstration projects and the S&T Program are expected to enter the ongoing program management decision process through the third component, a formal adaptive management process referred to as Adaptive Environmental Assessment and Management (AEAM; also with an estimated cost of $100 million). This chapter reviews each of these components of the LCA Study in turn, beginning with proposed on-the-ground restoration projects.

TABLE 6.1 LCA Study Estimated Restoration Costs

Item	Cost
Mississippi River Gulf Outlet environmental restoration features	$80,000,000
Small diversion at Hope Canal	$10,645,000
Barataria Basin barrier shoreline restoration	$181,000,000
Small Bayou Lafourche reintroduction	$75,280,000
Medium diversion with dedicated dredging at Myrtle Grove	$142,920,000
Subtotal	*$489,845,000*
Land easements, rights of way, relocation, and disposal	$178,619,000
First cost *Subtotal*	*$668,464,000*
Feasibility-level decision documents	$54,673,000
Preconstruction, engineering, and design	$36,252,000
Engineering and design	$29,018,000
Supervision and administration	$68,973,000
Project monitoring	$6,685,000

TABLE 6.1 Continued

Item		Cost
Conventionally authorized[a] cost	Subtotal	$864,065,000
Science and Technology Program (10 years)		$100,000,000
Demonstration program (10 years)[b]		$100,000,000
Beneficial use of dredge material program[b]		$100,000,000
Investigations of modifications of existing structures		$10,000,000
Authorized total LCA Study cost		**$1,174,065,000**
Multipurpose operation of Houma Navigation Canal Lock[c]		—
Terrebonne Basin barrier shoreline restoration		$84,850,000
Maintain land bridge between Caillou Lake and Gulf of Mexico		$41,000,000
Small diversion at Convent/Blind River		$28,564,000
Increase Amite River Diversion Canal influence by gapping banks		$2,855,000
Medium diversion at White's Ditch		$35,200,000
Stabilize Gulf shoreline at Point Au Fer Island		$32,000,000
Convey Atchafalaya River water to northern Terrebonne marshes		$132,200,000
Modification of Caernarvon diversion		$1,800,000
Modification of Davis Pond diversion		$1,800,000
	Subtotal	$360,269,000
Land easements, rights of way, relocation, and disposal		$208,100,000
First cost	Subtotal	$568,369,000
Feasibility-level decision documents		$47,529,000
Preconstruction, engineering, and design		$36,027,000
Engineering and design		$45,635,000
Supervision and administration		$58,673,000
Project monitoring		$5,683,000
Approved projects requiring future Congressional		
authorization for construction		**$761,916,000**
Mississippi River hydrodynamic study		$10,250,000
Mississippi River Delta management study		$15,350,000
Third Delta study		$15,290,000
Chenier Plain freshwater and sediment management and		
allocation reassessment study		$12,000,000
Acadiana Bays estuarine restoration feasibility study		$7,110,000
Upper Atchafalaya Basin study[d]		—
Large-scale and long-term studies cost	Subtotal	**$60,000,000**
Total LCA Study restoration cost		**$1,995,981,000**

[a]"Conventionally authorized" refers to items proposed in the Chief's Report and authorized by Congress through the Water Resources Development Act.

[b]Program total costs include any estimated real estate costs for these activities.

[c]Feature of the Mississippi River and Tributaries' Morganza, Louisiana, to the Gulf of Mexico Hurricane Protection Project.

[d]Study to be funded under the Mississippi River and Tributaries Authority.

SOURCE: U.S. Army Corps of Engineers, 2004a. (Refer to U.S. Army Corps of Engineers [2005b] for cost revisions.)

THE FIVE MAJOR RESTORATION FEATURES

Five near-term critical restoration features were selected utilizing the criteria discussed in the previous chapter. A brief description of each (excerpted from the LCA Study) follows. Although these features represent the most significant component of the restoration efforts evaluated in the LCA Study, they are still only the first stage of a long-term restoration program for coastal Louisiana.

Mississippi River Gulf Outlet

Discussed at length in Chapter 3, the Mississippi River Gulf Outlet (MRGO), originally defined as a restoration study during the framework selection process, appears in the final project description as consisting of the construction of 61.2 kilometers (km) (38 miles [mi]) of rock breakwaters to prevent high rates of erosion along the north bank of MRGO and at critical points along the southern shoreline of Lake Borgne that are in peril of breaching (U.S. Army Corps of Engineers, 2004a). This feature is expected to protect 25.7 square kilometers (km^2) (9.9 square miles [mi^2]) of marsh over the next 50 years at a cost of $108.3 million[1] (or $42,140 per hectare [$17,052 per acre]). It is interesting to note the recommended change for MRGO from closure in *Coast 2050: Toward a Sustainable Coastal Louisiana* (Coast 2050) to a study and then to repair in the LCA Study. Of the five features, the justification for MRGO is the most poorly documented and appears to be the weakest. In 2004, a concurrent resolution was passed by the Louisiana State House and Senate calling on the U.S. Army Corps of Engineers (USACE) to "promptly close" MRGO. At the time of this writing, some speculation about the role MRGO may have played in enhancing the storm surge during Katrina and the subsequent flooding of St. Bernard Parish was being put forward. What is known is that the levee that separates MRGO from St. Bernard Parish was topped during Katrina, as predicted in advance by some storm surge models.

Small Diversion at Hope Canal

The Hope Canal diversion includes construction of culverts in the Mississippi River levee and a receiving pond reinforced with riprap to slow flow and remove heavy sand, excavation of a new leveed channel from the existing south terminus of Hope Canal, enlargement of the Hope

---

[1]USACE, in the 2005 Chief's Report, updated the cost of the proposed MRGO feature to be $105.3 million (U.S. Army Corps of Engineers, 2005b).

Canal cross section, implementation of outfall management measures to direct water to the Maurepas Swamp, and installation of navigable constrictions. The feature is designed to restore approximately 145 km^2 (56 mi^2) of swamp at a cost of $70.5 million[2] (or $4,862 per hectare [$1,967 per acre]) (U.S. Army Corps of Engineers, 2004a).

Barataria Basin Barrier Shoreline Restoration

This feature involves dredging and placement of sand from Ship Shoal onto Caminada Headland and Shell Island (east and west), sediment nourishment and replanting of eroding marsh and dune, and removal of failing breakwaters. These activities are expected to increase dune and berm area by 2.6 km^2 (1 mi^2) and saline marsh area by 7.2 km^2 (2.8 mi^2) on Caminada Headland and to increase barrier island habitat by 0.6 km^2 (0.2 mi^2) on Shell Island over the next 50 years at a cost of $247.2 million[3] (or $237,692 per hectare [$96,187 per acre]) (U.S. Army Corps of Engineers, 2004a).

Small Bayou Lafourche Reintroduction

This feature will upgrade the existing sediment slurry pump and siphon facility to operate at the full capacity (9.6 cubic meters [m^3] per sec [12.9 cubic yards {yd^3} per sec]), construct a new 18.7 m^3 per sec (24.5 yd^3 per sec) pump facility, improve channel capacity by dredging and weir removal, stabilize the channel bank, operate water monitoring stations, install adjustable weirs at Thibodeaux and Donaldsonville to control water levels, and construct a sediment trap at Donaldsonville to control main channel siltation. The estimated cost is $144.1 million.[4] This feature proposes reliance on a heavily engineered approach (i.e., pumping sediment-laden water) with significant constructed features rather than recreating a more natural distributary channel. After 50 years, there would be approximately 10.1 km^2 (3.9 mi^2) more marsh (at a cost of $142,673 per hectare [$57,732 per acre]) in the project area than if the feature were not built (U.S. Army Corps of Engineers, 2004a).

---

[2]USACE, in the 2005 Chief's Report, updated the cost of the small diversion at Hope Canal to be $68.6 million (U.S. Army Corps of Engineers, 2005b).

[3]USACE, in the 2005 Chief's Report, updated the cost of the Barataria Basin shoreline restoration feature to be $242.6 million (U.S. Army Corps of Engineers, 2005b).

[4]USACE, in the 2005 Chief's Report, updated the cost of the small Bayou Lafourche reintroduction to be $133.5 million (U.S. Army Corps of Engineers, 2005b).

Medium Diversion with Dedicated Dredging at Myrtle Grove

Major elements of this feature are a gated diversion structure near Myrtle Grove with 141.6 m^3 per sec (185.2 yd^3 per sec) capacity, inflow, and outflow channels (4,877 meters [m] [16,000 feet {ft}]); channel guide levees and infrastructure relocation; and creation of 26.3 km^2 (10.2 mi^2) of new marsh through dedicated dredging. The estimated cost is $294 million[5] (or $111,787 per hectare [$45,238 per acre]). It is expected to prevent significant loss of intermediate, brackish, and saline marsh in the Barataria Basin (despite continued soil subsidence) and to improve sustainability of the Lafitte and Barataria communities and industries. This feature would support opportunities for demonstration projects and AEAM efforts (U.S. Army Corps of Engineers, 2004a).

OTHER ELEMENTS OF THE LCA STUDY

The Mississippi Valley, New Orleans District of USACE has the largest channel operations and maintenance program in USACE dredging 54 million m^3 (70.6 million yd^3) annually. More than 11 million m^3 (14.4 million yd^3) is used beneficially. The LCA Study requests authority for beneficial use of the sediment up to an additional 23 million m^3 (30.1 million yd^3) (U.S. Army Corps of Engineers, 2004a). USACE has also requested authority to investigate modification of existing structures, such as Davis Pond, Bonnet Carre Spillway, MRGO, Bayou Sorrel Lock, and Leland Bowman Lock (U.S. Army Corps of Engineers, 2004a).

Demonstration Projects

Five demonstration projects are included in the LCA Study. These projects are intended to support long-range planning and, in some cases, may represent precursors to larger-scale future projects. Some are synergistic with major projects or their components. The generic characteristics of these projects encompass the following: (1) marsh restoration and/or creation using nonnative sediment, (2) marsh restoration using long-distance conveyance of sediment, (3) canal restoration using different methods, (4) shoreline erosion prevention using different methods, and (5) barrier island restoration using offshore and riverine sources of sediment.

---

[5]USACE, in the 2005 Chief's Report, updated the cost of the medium diversion at Myrtle Grove to be $278.3 million (U.S. Army Corps of Engineers, 2005b).

Science and Technology Program

The goal of the S&T Program is to provide a sound basis to effectively address coastal ecosystem restoration needs. It is intended to improve coastal restoration decision making; provide scientific data, analysis, and interpretation; develop tools, methods, and protocols for restoration planning and assessment; minimize knowledge gaps that limit restoration planning and execution; assess the effectiveness of restoration actions in meeting LCA Study goals; and provide information and synthesis in a rapid and useful manner. Key tools to accomplish these goals are considered to be sound baseline data, monitoring over time and space, models, data management, and continued research. Cyber-infrastructure will have to be a key component of the S&T Program to facilitate access to information and communication among agencies, universities, other interested parties, and the general public. The program management called for in the LCA Study includes independent technical review committees and advisory boards and reviews of existing data through conferences and meetings (Figure 4.5) (U.S. Army Corps of Engineers, 2004a). An annual review and update is planned. The S&T Program, as described in the LCA Study, focuses on the physical and natural sciences. Although this is appropriate, the impacts on the human population are also important. Therefore, economic, urban planning, sociological, and public policy expertise in the S&T Program appears to be insufficient.

There is an inherent conflict between the required independence of the S&T office from the LCA Study management and the need for significant involvement and insertion of S&T results into LCA Study decision making that is critical for adaptive management. Since the skills needed for this type of interdisciplinary work are not readily available, LCA Study planners have to acknowledge and strive to resolve or, at least, reduce the impact of this conflict.

Modeling

A key role for the S&T office is development of analytical tools. This includes development, revision, and refinement of hydrodynamic and ecosystem modeling, which is considered fundamental to the S&T charge. Appendixes A and C of the LCA Study (U.S. Army Corps of Engineers, 2004a) outline the key roles of models and the different types of mathematical models—natural and resource, engineering, and economic. (See Chapter 5 for detailed discussion and analysis of the modeling used to develop the LCA Study.)

Peer Review

Peer review of program planning is considered a key element; yet, it is unclear who will identify and select peer reviewers and what pool of candidates might be considered (U.S. Army Corps of Engineers, 2004a). As recommended in the National Research Council (2002) report *Review Procedures for Water Resources Project Planning*, complex water resources project planning studies undertaken by USACE should be subject to external, independent review. One or more panels of impartial, highly qualified experts should conduct this external review. External review panels should not include USACE staff members and should not be selected by USACE. External reviews should be overseen by an organization independent of USACE, which will provide the highest degree of credibility of review (National Research Council, 2002).

Uncertainties

A primary role of the S&T office is the identification of uncertainties and funding for projects to reduce these uncertainties. The LCA Study (U.S. Army Corps of Engineers, 2004a) describes these knowledge gaps as "uncertainties" and lists the following four types:

• Type 1: Physical, chemical, geological, and biological baseline conditions
• Type 2: Engineering concepts and operational methods
• Type 3: Ecological processes, analytical tools, and ecosystem response
• Type 4: Socioeconomic and political conditions and responses

These knowledge gaps are real, and the proposed demonstration projects to alleviate these deficiencies are definitely valid. However, the LCA Study fails to acknowledge several other important knowledge gaps, which are discussed at some length in Chapter 7. More specific guidance to address uncertainties or knowledge gaps is also presented in Chapter 7; however, it is appropriate in this discussion of the interactions with various components of the S&T Program to point out that the LCA Study appears to place insufficient emphasis on some key strategies for reducing uncertainties of this type: (1) use of existing literature and information; (2) use of available but uncollated and unsynthesized data; (3) professional experience; (4) bench-, micro-, and mesocosm-scale studies; (5) expansion of existing projects; (6) field trials using intermediate-scale demonstrations; (7) prototype-scale demonstrations; (8) undertaking these analyses within a formal adaptive management process; and (9) application of numerical models.

ADAPTIVE MANAGEMENT

The adaptive management process for the LCA Study is referred to as AEAM (U.S. Army Corps of Engineers, 2004a). The theory, importance, and essential elements of a successful adaptive management process for natural resources management are well documented in *Adaptive Management for Water Resources Project Planning* (National Research Council, 2004b). The six identified elements of adaptive management are as follows:

1.   Management objectives that are routinely reviewed and revised, as needed
2.   Management system model(s)
3.   A variety of management choices
4.   The monitoring and evaluating of outcomes
5.   Mechanism(s) to include lessons learned in future decisions
6.   A collaborative method for stakeholder involvement and education

The underlying premises of the AEAM process are that (1) knowledge gaps exist when attempting to manipulate a large-scale ecosystem and (2) a need exists for an iterative and flexible approach to management and decision making. Therefore, the AEAM process is designed to allow for future actions to be changed based on observing the efficacy of past actions on the ecosystem. AEAM is also important in dealing with wicked problems (see Chapter 5), wicked unknowns, and decision-making relationships. The AEAM process is described in the LCA Study as supporting *passive* adaptive management, which currently occurs with Coastal Wetlands Planning, Protection, and Restoration Act (CWPPRA) projects, and *active* adaptive management, as used with the Caernarvon Freshwater Diversion.

The proposed AEAM process indicates that all organizations within the LCA Study's management structure have a role in implementing AEAM; however, the S&T office will have primary responsibility for making recommendations to the program management team (PMT) and the program execution team, based on the assessment of monitoring data and the development of new technologies (U.S. Army Corps of Engineers, 2004a). These two teams then formulate and implement any needed adjustments to the program or projects.

The AEAM process is proposed as a three-phase cycle consisting of (1) decision making by PMT; (2) implementation by the program execution team (physical and operational); and (3) monitoring, assessment, and reporting by the S&T office. A review of the AEAM process indicates an appropriate effort to integrate sound science in understanding the efficacy of past actions on the ecosystem in order to modify or change future actions. Effective adaptive management is, and will continue to be, critical to the long-term success of the Louisiana restoration program.

It is not clear what mechanisms are in place to incorporate the "learning" aspect of monitoring and assessment outcomes. Although the role of science (especially monitoring and assessment) is a central principle of adaptive management, an effective adaptive management process is much more than just the integration of good science. Adaptive management is a process that requires the integration of learning and adaptation throughout all aspects of decision making and implementation (National Research Council, 2004b).

To successfully evaluate the efficacy of Louisiana's restoration activities, objectives must be developed for the overall program, as well as for individual projects. These objectives have to be clear, specific, quantifiable, and at a scale appropriate to their purpose. Of particular importance are project-level objectives against which monitoring outcomes will be assessed. The absence of such objectives creates knowledge gaps and ambiguity as to what future adjustments need to be made, as well as to the success of the restoration effort.

AEAM will depend heavily on routine and continuous monitoring, numerical model application and data, and information management and communication. (See U.S. Army Corps of Engineers' [2004a] Appendix A for details.) These activities are envisioned as taking place entirely within the domain of Louisiana's coastal area. There is no acknowledgment that major national investments are being made and will soon be accelerated in connection with the coastal ocean observing and watershed monitoring programs. Opportunities currently exist for the coastal Louisiana restoration to coordinate activities with ongoing environmental data collection efforts to avoid wasteful duplications and take advantage of resources that will be funded from outside the restoration budget. If the information that emerges from restoration observations and model output is to be credible and utilized effectively by others (thus maximizing the benefit of data collection), it should comply with a set of standards and information communication protocols that are now in an intermediate stage of development under the auspices of the data management and communication working group within the Ocean.US integrated and sustained ocean observing system.

Finally, one of the essential elements required in a successful adaptive management process is the inclusion of a collaborative structure for stakeholder participation and learning. Stakeholder participation is missing from the AEAM process, and meaningful stakeholder involvement seems to be absent in the overall implementation of the LCA Study. As stated by the National Research Council (2004b), "achieving meaningful stakeholder involvement that includes give and take, active learning (through cooperation with scientists), and some level of agreement among participants, represents a challenge but *is essential to adaptive management*" (emphasis added). Stakeholder participation needs to be part of AEAM and should include, at a minimum, representatives from local government, industry, key organizations, and citizens at large.

PROPOSED MANAGEMENT APPROACHES

The LCA Study presents the proposed management structure of the restoration program and describes the roles of the various entities involved in the implementation of this program (Figure 4.5). What is more important is a description of *what* the entities will do; this description lacks crucial information regarding the institutional mechanisms of *how* they will do it. The proposed decision support system is supposed to overcome this problem. "Management of the [Louisiana] restoration efforts would also include a decision support system that relies on clearly defined procedures to assess knowledge gaps and develop alternatives for the decision-making process" (U.S. Army Corps of Engineers, 2004a). The effectiveness of the decision support system will be one of the determining factors in the success of the restoration program, and its development should be peer-reviewed prior to adoption. A review of the proposed management structure and of the description of roles reveals several areas in which further clarification and modification are appropriate.

The organizational structure includes a recommendation for the U.S. Congress to authorize the establishment of a task force whose purpose would be to "facilitate coordination and collaboration among various agencies involved in implementation of major coastal restoration activities and provide recommendations to the Secretary of the Army" (U.S. Army Corps of Engineers, 2004a). Establishment of such a task force would be beneficial to the restoration program by ensuring senior-level support among the key participants in the program; however, the benefit of having a regional working group separate from PMT is not clear. It would be more efficient, ensure better participation and commitment, and strengthen linkages between the senior agency officials and the restoration program for PMT to be expanded to include the regional federal agency representatives.

The LCA Study's management strategies rely heavily on the interaction of numerous groups and committees that will provide direction, assessment, and feedback to the program through an adaptive management process. The success of implementing an effective adaptive management process will rely heavily on day-to-day lines of communication, resolute decision making, and constant follow-through by the program manager who will have overall responsibility for the restoration program.

The management structure proposed in the LCA Study does not include any formal opportunity for the public or stakeholders to have input to decision making. This is an issue that goes hand-in-hand with deficiencies in the adaptive management program discussed earlier in this chapter. The management plan needs to better clarify the decision-making process, the points of entry for public input, and the mechanisms for stakeholder involvement.

FEASIBILITY

The feasibility of the LCA Study is analyzed from a number of perspectives, including sediment transport, hydrologic, ecological, economic, engineering, and cultural and social considerations.

Riverine Sediment Transport

Sedimentation is arguably the most fundamental of all processes that control the potential for sustaining existing areas or building new wetlands. The spatial distribution of sedimentation and the associated quantities will govern the future landscape within coastal Louisiana and determine those regions that will be sustainable or will be lost to subtidal habitat. Despite this key role, considerable uncertainty and differing professional opinions exist on the quantities of sediment available and how easily they can be transported to the desired location. Without a clear understanding of the expected annual delivery of sediments of a particle size capable of deposition under different conditions, it is difficult to assess the area of wetlands that can be sustained or built.

Average Annual Sediment Delivery to the Lower Mississippi

Several investigators have reported a large reduction in the sediment load of the lower Mississippi River since 1893 (Robbins, 1977; Keown et al., 1981; Kesel, 1988, 1989, 2003) (Figure 2.2). The anthropogenic changes to the channel that influence the quantities of sediment being transported through the lower river include (1) steepening of the longitudinal profile by eliminating meanders; (2) bank stabilization to reduce bank erosion; (3) capture of instream sediment by dike fields; (4) construction of levees that eliminate most flooding and sediment distribution across the marsh plains, particularly through crevasse splay (overbank flooding and associated deposition of sediments); (5) construction of dams in the Mississippi River Basin interrupting the flux of sediment through the watershed; and (6) changing land use throughout the basin.

Changes 1 and 4 tend to increase local sediment transport capacity and discharge to the Gulf of Mexico; changes 2, 3, and 5 reduce sediment transport overall, and change 6 could increase or decrease sediment transport. Harmar (2004) has shown that degradation of the channel (implying scour due to elimination of meanders) minimizes the flow of the Old River, which flows from the Mississippi River to the Red and Atchafalaya Rivers.

Table 2.1 provides an estimate of the overall changes in suspended sediment passing New Orleans, where the reduction in sediment load prior to 1950 was attributed primarily to decreased river discharges and land-use changes throughout the basin. The reductions in the 1950s and 1960s were attributed to the completion of major dams on the Missouri and Arkansas Rivers. There are few detailed measurements of bedload (sand particles moving close to the bed, and sand waves propagating along the bottom of the channel), but it is thought to be approximately 30 percent of the suspended sediment load (Kesel et al., 1992; Kesel, 2003).

As the Mississippi River was channelized with levees, revetments, and dikes and the upper watershed was dammed, both the upstream sediment sources and the availability of sediment from lateral erosion diminished. For example, Kesel (2003) estimates that 90 percent of overbank flooding in the lower Mississippi River had been eliminated by 1927. There has been a significant reduction in lateral migration of the channels and no major crevasses formed due to the heavily engineered levees. Biedenharn (1995), Biedenharn et al. (2000), and Thorne et al. (2000), however, provide a different analytical approach related to sediment availability (see Chapter 2).

Compounding the difficulties of how much sediment might be available for land building or replenishment under different management actions is the potential role of washload in marsh development. Also important to the landscape-scale changes in Louisiana's coastal area is how the sediment will be distributed. The changes in the lower Mississippi River have eliminated most of the pre-channelization pathways that used to help sustain marsh elevations (Figure 6.1; Kesel, 2003).

Vertical Sediment Concentration Profiles and Recruitment of Organic Component

There is uncertainty regarding the vertical distribution of suspended sediment over the depth of the river (often referred to as the vertical suspended sediment concentration profile). Wells (1980) and Kesel (1988) report data indicating that the concentration is uniformly distributed over

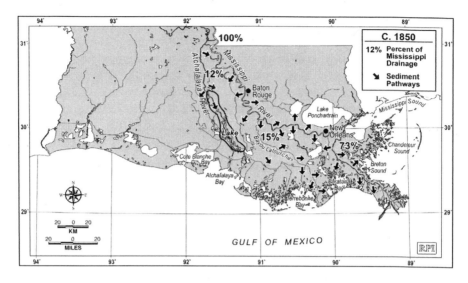

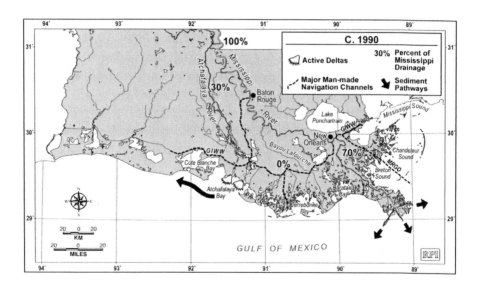

FIGURE 6.1 Comparison of the percentage of Mississippi River discharge flowing into the adjacent wetlands and distributary channels on the Louisiana deltaic plain for 1850 and 1990 (modified from Kesel, 2003; background maps supplied by Research Planning, Inc.). The arrows in the lower panel are intended to indicate the general direction, and not magnitude, of sediment transfer onto the shelf. NOTE: GIWW refers to the Gulf Intracoastal Waterway.

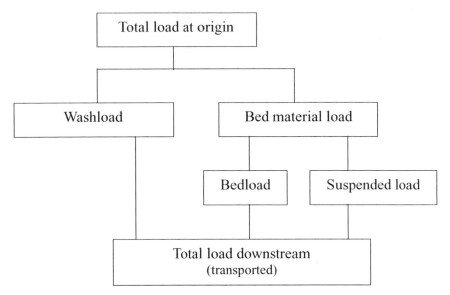

FIGURE 6.2 Definitions of sediments being transported in a river. Bed material is the sediment mixture of which a streambed, lake, pond, or estuary bottom is composed. Bedload transport is material carried by the river in near-continuous contact with its bed. Suspended load is the material that is transported in the water column and is maintained in suspension through the turbulence in the water. The washload is the portion of the total sediment load composed of grain sizes finer than those found in appreciable quantities in the streambed (modified from Thorne et al., 2000).

the depth, but other interpretations indicate higher concentrations close to the bed (Figure 6.2).

Marsh soils contain a high percentage of organic material; thus, the soil horizon consists of both organic and inorganic components. The recruitment of the organic component and the rates of decomposition are discussed in Chapter 2.

Hydrology

The cumulative effects of regulated flows through the Mississippi River Basin influence the delta region. Flow regulation results in less intra-annual variability, although episodic extremes will still occur as evidenced by recent floods and droughts. The flows are important in several physical processes that govern the sustainability of marsh regions, such as the ability to transport sediment (refer to the following section) in the channels and on marshplains and the control of salinity levels, rates of organic material recruitment, and ecological viability.

Ecological Considerations

A series of ecological considerations played a limited role early in the LCA Study's project selection. Ecological services evaluated include land building, habitat switching, primary productivity, nitrogen removal from the Mississippi River, and habitat use. (See Chapter 5 for details.) However, the large uncertainties in the model predictions of ecological services (U.S. Army Corps of Engineers' [2004a] Appendix A) make it even more important to have solid, ongoing monitoring that is the basis for adaptive management.

Economic Assessment

The LCA Study briefly discusses the ecologic and economic significance of coastal Louisiana to the nation (U.S. Army Corps of Engineers, 2004a). This significance includes the major deep draft ports located in southern Louisiana, the offshore continental shelf oil and gas production, the movement of oil and gas through Louisiana, the large population living in coastal Louisiana, the large volume of commercial fisheries, the large recreational expenditures, and the high value of capital infrastructure located in coastal Louisiana. Since many people depend on coastal Louisiana for their well-being, it is clear that it is a major center of U.S. economic activity. Therefore, alterations in ecological conditions of coastal wetlands *may* have substantial impacts on national welfare. However, two primary questions must be addressed:

1. Have USACE and Louisiana, in the LCA Study and in supporting reports, established the national significance of the coastal Louisiana wetlands and barrier island ecosystems?
2. Have reasonable procedures been employed in selecting the array of projects for the LCA Study to address wetlands and barrier island loss?

The LCA Study suggests that coastal Louisiana ecosystem services are significant to the nation, based on "counts," such as how much oil and gas pass through Louisiana, how many fish are caught, how many people recreate, and how much infrastructure is at risk. At a more meaningful level of what "difference" these services make to the national economy, the LCA Study and its supporting documents did not make a compelling case that saving coastal Louisiana will make a substantial difference to the national economic well-being.

While the magnitude of economic activity dependent on coastal Louisiana ecosystems suggests a high level of significance, further reflection makes it much less clear. The LCA Study presents sufficient information about the importance of some components of the natural and built environment in coastal Louisiana (e.g., system of deep water ports, oil and gas receiving and transmission facilities, complex and extensive urban landscape, robust commercial fishery) to demonstrate that substantial economic interests are at stake in coastal Louisiana and that these interests have national significance. The immediate impacts of Katrina underscore the important role that New Orleans and adjacent areas of the Gulf Coast play in the national economy.

Establishing the true, national economic significance of efforts to restore coastal wetlands in Louisiana as proposed in the LCA Study, however, must go beyond simply identifying and characterizing these components and should (1) include an analysis of how specific restoration efforts will preserve or enhance their value (i.e., some restoration efforts may have little influence on the vulnerabilities of specific components of natural and built environment in coastal Louisiana) and (2) determine how the national economy would respond to the loss or degradation of these components (e.g., what is the capacity for similar components in other regions to compensate for the loss and on what time scales?).

Considering only the significance of oil and gas as an example, less and less of Louisiana's production will be at risk from coastal degradation over time as the at-risk fields are being depleted. The offshore continental shelf oil and gas industry dependence on coastal Louisiana is high under current locations of service activities and pipelines. However, onshore infrastructure can be relocated over time as those facilities depreciate, and this may be less expensive than trying to save the areas where dependence is high. Shipping vulnerabilities to coastal degradation can be remedied by complete reconfigurations of routes, ports, and transportation modes. The large economic value of coastal infrastructure, residences, businesses, and transportation networks is not immutable and could be scaled down under coastal relocation policies and may be less expensive to abandon than to protect. There are considerable knowledge gaps as to the long-term impact that restoration will have on the oil and gas industry.

As a greater understanding is achieved of the short- and long-term economic impacts of Katrina, a more meaningful effort to evaluate the national economic significance of protecting the natural and built environment in coastal Louisiana will be possible. Such information would provide an important context for decision making; however, it will still be important to understand the role wetlands play in protecting specific components of the overall system and to determine how specific restoration efforts can enhance that protection. While wetlands and adjacent barrier islands and levees are known to reduce impacts from waves, their more complex role in reducing storm surge is less understood. Surges contain multiple components, including barometric tide effects, wind stress-induced setup, wave-induced setup, and Coriolis forces. As was pointed out repeatedly in the public media during Katrina and Rita, in the northern hemisphere, the eastern side of a hurricane tends to drive water northward in a counterclockwise manner. If a storm stalls off a coast for a significant period of time, it will continue to drive water onshore for a prolonged period, regardless of the nature of any intervening wetland or barrier island. Thus, the potential for reducing risk due to storm surge is more difficult to generalize.

Overall, the information necessary to fully understand the economic implications of wetland loss does not currently exist. Systematic studies of the impacts of hurricanes, such as Katrina and Rita, would be helpful, but even then a specific end state of restoration effort would have to be identified so that a projected distribution of wetland would be available for any analysis of risk.

Engineering Approach

The Mississippi River Delta is part of a multifaceted social, regulatory, economic, and political environment and is an immensely complex engineered system. The response of the delta system to anthropogenic activity emphasizes the necessity of analyzing and planning in a comprehensive manner because actions taken at one location will have implications elsewhere. The levees and large control structures, such as the Old River Control Structure and the Old River Auxiliary Control Structure, limit the flow of water and sediment through the Atchafalaya River and adjoining wetlands. It is assumed that this condition can be sustained for the life of the project. Reduction of annual sediment loading to the delta from historic levels (averaging about 450 million metric tons [992 billion pounds] in 1880 to less than 150 million metric tons [331 billion pounds] in 1980) is exacerbated by subsidence, eustatic sea level rise, consolidation, and oxidation. However, other investigators have drawn different conclusions about changes in the sediment delivery to the coastal zone (discussed in Chapter 2 and in this chapter), and this remains an important uncertainty. Sediment available for restoration is a finite resource. Previous investigators have identified several sources of sediment, including upstream sources, the bed of the channel, and the banks. Upstream sources have been reduced due to dams, and bank sources have been reduced by the stabilization of the levees and banks. If channel deepening has been a major source of sediment, which is diminishing as the channel approaches some new dynamic equilibrium (or if large new diversions reduce downstream flows), there is a real possibility that less sediment will be available for restoration. This means there is slower recovery or fewer areas that can be restored.

Cultural and Social Aspects

The human component of the study area includes large metropolitan areas around New Orleans (comprising Jefferson, Orleans, Plaquemines, St. Bernard, St. Charles, St. James, St. John the Baptist, and St. Tammany Parishes) with a population of 1.3 million in 2000 and small communities sprinkled throughout the rest of the landscape. The area's population is growing, with fewer people per household. The implication, however, is that more land is required to accommodate potentially significant population increases in areas where land area is diminishing (U.S. Army Corps of Engineers, 2003a). Community and regional growth would not have been possible without construction of an extensive network of levees and floodgates along the Mississippi River and numerous lesser flood control and hurricane protection projects for flood protection.

Historically, most of the activities that have driven regional and community growth have centered on oil and gas production, tourism, port operations, fishing, and hunting. Development of the area's energy resources during the 1950s and 1960s was instrumental in the expansion of industrial growth in the surrounding communities. More recently, saltwater sport fishing has become an important stimulus to local and regional economies (U.S. Army Corps of Engineers, 2003a). These activities are the impetus for the projects, but as pointed out above, purely economic analysis may suggest that abandonment of some towns and industrial sites is more cost-effective from a national perspective.

SOME CONSIDERATIONS FOR LONG-TERM PROJECTS

Over the long term, it will be essential to plan for larger-scale and bolder projects than those proposed in the near-term LCA Study. The total LCA Study's budget of $1.9 billion includes $60 million for studies of potential large-scale and long-range projects. Two projects that should receive serious further study include the Third Delta and the complete abandonment of the active Birdsfoot Delta lobe. The LCA Study identifies both the Third Delta study and Mississippi River Delta management as large-scale, long-term concepts requiring detailed study; however, no consideration is offered for evaluating abandonment of the active Birdsfoot Delta.

Third Delta

The Third Delta envisioned in the LCA Study would divert water from the Mississippi River near Donaldsville through a 88.5-km (55-mi) long conveyance channel to the Barataria and Terrebonne Basins (Regions 2 and 3) as shown in Figure 6.3. Two diversion designs with maximum river discharges of 3,398 m^3 per sec (4,444 yd^3 per sec) and 6,796 m^3 per sec (8,889 yd^3 per sec) are envisioned in the LCA Study. Examined below are the present and projected land loss rates, the wetland formation benefits of a Third Delta, and the anticipated difficulties.

The projected land changes resulting from current CWPPRA and LCA Study projects over the next 50 years are summarized in Table 6.2 (U.S. Army Corps of Engineers, 2004a). If the land gain and loss estimates are correct, there will be an average net land loss of 26.7 km^2 per yr (10.3 mi^2 per yr) without construction and implementation of the LCA Study's five recommended restoration features. Implementing these restoration efforts would reduce this land loss by 4.4 km^2 per yr (1.7 mi^2 per yr) to 22.3 km^2 per yr (8.6 mi^2 per yr) at a cost of approximately $20 million per km^2 ($51 million per mi^2) of reduced loss over this 10-year period; however, the wetland benefits (gains or loss prevention) are projected to continue over 50 years, thereby reducing the cost per unit area. At these unit costs, the annual cost required to eliminate net land losses would be approximately $525 million using 10-year rates or $105 million using 50-year rates. "The goal of [the projects proposed in] the LCA [Study] is to reverse the current trend of degradation of the coastal ecosystem" (U.S. Army Corps of Engineers, 2004a), which will require reducing unit costs of land gain and loss prevention to the degree possible. The Third Delta is one means of potentially reducing the unit costs through the delivery of large volumes of sediment and the associated economies of scale.

To evaluate the efficacy of the Third Delta in sediment delivery and relative wetland gain, it is useful to develop very approximate conversion factors between a unit of water diverted and a unit area of wetland gain. This can be based on the combined average discharges of the Mississippi and Atchafalaya Rivers and their associated mineral sediment loads, as summarized in Table 6.3, and on the approximation that one unit of mineral sediment will result in 4–10 units of wetland volume and that the average effective vertical dimension of wetland generated is 2 m (6.6 ft). These conversion ratios are shown in Table 6.4.

Thus, it appears that within the assumptions and considerations applied, implementation of the Third Delta project could more than offset projected net losses; however, the resulting gain would not occur in the same locations or possibly with the same wetland quality as the losses. As noted, the Third Delta as planned would not contribute to the reinstate-

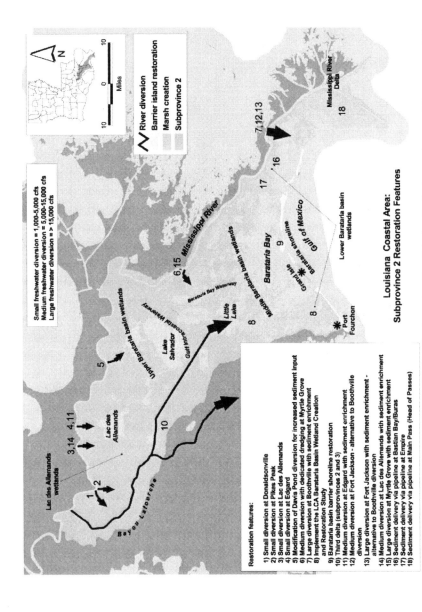

FIGURE 6.3 Various restoration features considered, including the Third Delta (U.S. Army Corps of Engineers, 2004a; used with permission from the U.S. Army Corps of Engineers). (NOTE: cfs = cubic feet per sec.)

TABLE 6.2 Land Change Projections Under Various Considerations

Consideration	Land Changes in km² per yr (mi² per yr in parentheses)
Loss reduction due to funded CWPPRA and other projects	6.7 (2.6)
Existing situation with funded CWPPRA and other projects	−26.7 (−10.3)
Anticipated reduction due to the LCA Study	4.4 (1.7)
Net change with the LCA Study	−22.3 (−8.6)

SOURCE: Modified from U.S. Army Corps of Engineers, 2004a.

TABLE 6.3 Estimates of Average Annual River Flows and Sediment Delivery

Combined Average Mississippi River and Atchafalaya River Flows in m³ per sec (yd³ per sec in parentheses)	Combined Mineral Sediment Delivery in m³ per sec (yd³ per sec in parentheses)
19,737 (25,815)	4 (5)

SOURCE: Data from U.S. Geological Survey, 2005.

TABLE 6.4 Wetland Gain Based on Third Delta Diversion Discharge Rates

Design Maximum Discharge in m³ per sec (yd³ per sec in parentheses)	Ratio (R) of Mineral Sediment to Wetland Volume	
	Annual Wetland Gain/Loss Prevention in km² per yr (mi² per yr in parentheses)	
	R = 1:4	R = 1:10
3,398 (4,444)	10.9 (4.2)	28.0 (10.8)
6,796 (8,889)	22.3 (8.6)	55.9 (21.6)

NOTE: These calculations are based on the average diversion discharges of one-quarter the maxima. For a ratio of 1:4 mineral sediment to wetland gain, 1 yd³ per sec = 0.0034 acres per yr = 0.00014 mi² per yr, and for a ratio of 1:10 mineral sediment to wetland gain, 1 yd³ per sec = 0.0085 acres per yr = 0.00036 mi² per yr.

ment or maintenance of the barrier islands since most of the sediment would be finer than required for barrier island purposes. The barrier island restoration in the first five LCA Study projects may provide some of the sediment for barrier island restoration at the lower end of the Third Delta (Figure 6.3).

Anticipated Third Delta Challenges

The Third Delta challenges are broad and include engineering, ecological, and social dimensions. Here, the emphasis is on the stakeholders with issues of real estate (including obtaining right of way) and habitat, each of which could delay the Third Delta project indefinitely. The project, as described in the LCA Study, includes the construction of a 88.5-km (55-mi) long conveyance channel through both public and private property and the discharge of freshwater into Barataria and Terrebonne Basins. Securing right of way and other real estate issues could be costly and present political challenges, each of which could prove to be insurmountable. Despite USACE's power to take possession before questions of compensation have been settled, real estate issues could take decades to resolve with many costly and lengthy legal and political challenges. Although the Third Delta design may provide an overall benefit for stakeholders, some will recognize early benefits and some either will be impacted adversely or will perceive or claim this to be the case.

If initial investigations indicate that insurmountable difficulties exist with implementation of the Third Delta as now planned, it is suggested that an alternate plan be considered, such as a diversion much farther south, leaving a "slack water" channel gulfward. This option and the associated merits and challenges are discussed in more detail in Chapter 7.

Abandonment of the Active Birdsfoot Delta

Another alternative that has not yet received serious consideration or assessment by the LCA Study is the complete abandonment of the Birdsfoot Delta. As noted in Chapter 2, coastal dispersal of Mississippi River sediment has been dominated more by the processes of episodic avulsion of deltaic lobes than by ocean forcings (Wright and Nittrouer, 1995). This is of fundamental importance to understanding how the Louisiana coast differs from other deltaic coasts and is attributable in large measure to the generally low-energy ocean conditions that prevail except during storm events (Wright, 1995). Recent field studies on the Louisiana inner continental shelf have shown that fair weather conditions are incapable of resuspending and transporting recently deposited muds most of the time (Wright et al., 1997). For this reason, historically, elongated and relatively delicate delta lobes have extended well across the shelf before being abandoned and ultimately destroyed by storm waves.

Today, however, the prevalence of low-energy conditions is only partially responsible for the anomalous Birdsfoot Delta that crosses the entire Louisiana shelf before discharging its (potentially) coast-nourishing sediment load near the shelf break. Human intervention has reinforced this outcome. Not only does this impede the along-shelf transport system, it deprives the coastal areas to the west of land-replenishing sediment, including marsh-building muds, as well as fine sands that could nourish barrier islands. If the sediment that presently enters the Gulf of Mexico via Southwest Pass, South Pass, and Pass al'Outre were to enter further to the west in shallower waters, the sediment would be delivered where it would be more effective for wetland formation. In addition, the eventual disappearance of the Birdsfoot Delta protrusion would lead to a restructuring of the shelf physical ocean conditions and probably to a moderate strengthening of east-to-west flows, which as noted in Chapter 2 are presently separated in two distinct cells (Smith and Jacobs, 2005).

An alternative scenario for the retention of sand and silt now lost beyond the shelf break would involve diverting the main flow of the Mississippi River toward the west of its present main channel somewhere between New Orleans and Head of Passes. An intermediate- and long-term consequence of this action would be the abandonment of the active Birdsfoot Delta by the Mississippi River. A clear benefit would be the nourishment of eroding coastal reaches to the west. Although this alternative has been widely acknowledged as possible, its feasibility has, for various reasons, not been seriously considered by USACE. Therefore, unlike the Third Delta scenario, it is not yet possible to assess the potential advantages and disadvantages of Birdsfoot Delta abandonment. Obviously, implementation of such a strategy would have to be accompanied by the creation of a deep navigation access channel somewhere downstream of New Orleans but upstream of Head of Passes. Although the size of the area it would impact would make it controversial, such a diversion represents the type of project that deserves greater consideration as restoration efforts move forward.

ENHANCING THE FEASIBILITY OF THE OVERALL APPROACH

An undertaking as complex as restoring and protecting coastal Louisiana presents several challenges to the understanding of many natural and anthropogenic processes, as well as fundamental engineering design. Thus, a robust program is needed to identify key knowledge gaps as they are identified and to implement research and demonstration projects to provide the basis for effective problem solving. Analysis of the LCA Study suggests that major knowledge gaps exist that may require acknowledgment and addressing early in any implementation effort. Uncertainties associated with relative sea level change, bathymetric and topographic data quality, adequacy of resources, and distribution methods suggest that greater emphasis may have to be placed on environmental monitoring to provide adequate baseline information. In addition, process-based models for prediction of land building, which link socioeconomic outcomes to biophysical processes, will likely also be needed.

As pointed out in Chapter 5, models often play a key role in project selection and adaptive management. A modeling effort as complex as the one needed for restoring the Louisiana coast requires a great deal of observational data and modeling coordination. In addition, the best modeling is done when the process is open to collaborating scientists (physical, biological, and social) nationally and from around the world. A new modeling program, involving multiple state and federal agencies and academic institutions, is needed to draw together experience creating an open scientific forum. This forum will generate the confidence of scientists, engineers, and the public. The cost of developing such a program will be insignificant compared to the implementation costs, the scale and complexity, and the long-term nature of Louisiana's restoration process. Guidance developed through this program can be expected to reduce the risk of failures, anticipate future problems (physical, biological, social, political, and economic), and reduce the overall costs as experience and data result in improved projections.

The coastal Louisiana restoration endeavor would benefit greatly from a coordinated effort that would result from the establishment of a more robust informatics and modeling effort focusing on the lower Mississippi River. This effort should have the following characteristics:

- Leadership by a prominent academic or research scientist and a consortium of universities within the state, state and federal agencies, and federal research facilities
- Acceptance of new technologies and facilitation of open-access, community-wide scientific research on a national and global scale
- New technologies and open research with the national and global scientific community
- A scientific steering committee of internationally recognized modelers and scientists
- Facilitation of capacity building to train and educate (in disciplinary and interdisciplinary fashion) the next generation of engineers and scientists who will inherit the management of the lower Mississippi River

- University campus location but with resident representatives of all agencies involved in the restoration and management of the lower Mississippi River
- Cyber-infrastructure for communication, modeling (physical, biological, social, and economic), and data management for the restoration process
- Cyber-infrastructure that includes (1) the data warehouse, data archiving, computational core capable of running large models, and communication facilities for collaborating with national and local researchers and managers and (2) a web portal for the general public to track current conditions and various alternatives being investigated

The S&T Program envisioned in the LCA Study is an innovative and essential element that provides a process for planning and assimilating monitoring results and developing adaptive management strategies. In addition, the S&T Program is an appropriate administrative home for model development and maintenance. The proposed S&T Program represents a very positive step in the development of a coordinated approach for an adequate knowledge base of how coastal Louisiana may respond to various restoration efforts or evolve in the absence of some of those efforts. **The S&T Program requires a more explicit statement of program responsibilities and means for setting priorities; it must be integrated more effectively into the central management structure through the adaptive management process and include better representation of social sciences and ecological processes.**

It is unreasonable to expect any region to have all the necessary experience and human resources to address most effectively the challenges of the magnitude represented by land loss in coastal Louisiana. Just as the funding of the LCA Study and its extensions includes a combination of state and federal resources, the scientific and other elements of the LCA Study should draw on the best state, national, and international talents available. Therefore, **the LCA Study should direct efforts toward capacity building that enables the program to address its stated objectives by drawing on the widest possible pool of national and international technical expertise.**

The AEAM process for the LCA Study lays out a mechanism to integrate emerging technical information into the management processes, in addition to the use of sound science in understanding the efficacy of past actions in order to modify or change future actions. AEAM efforts, however, appear to be marginalized within the overall management structure. The full benefit of taking an adaptive management approach to complex projects, such as proposed in the LCA Study, will be realized only if it represents a core theme of the overall project management strategy. **Steps** should be taken to strengthen the AEAM process throughout the management structure.

CWPPRA, Coast 2050, and efforts to develop *Louisiana Coastal Area, LA—Ecosystem Restoration: Comprehensive Coastwide Ecosystem Restoration Study* (draft LCA Comprehensive Study) placed greater emphasis on stakeholder involvement than is evident in the LCA Study. **Stakeholder participation (including, at a minimum, representation from local government, industry, key organizations, and citizens at large) should be accounted for in the management structure of the Louisiana coastal area program.**

The LCA Study presents sufficient information about the importance of some components of the natural and built environment in coastal Louisiana (e.g., system of deep water ports, oil and gas receiving and transmission facilities, complex and extensive urban landscape, robust commercial fishery) to suggest that substantial economic interests are at stake in coastal Louisiana and that these interests have national significance. The immediate impacts of Katrina underscore the importance of New Orleans and adjacent areas of the Gulf Coast to the national economy. Establishing the true, national economic significance of efforts to restore coastal wetlands in Louisiana as proposed in the LCA Study, however, must go beyond simply identifying and characterizing these components and should include an analysis of how specific restoration efforts will preserve or enhance their value (i.e., some restoration efforts may have little influence on the vulnerabilities of specific components of the natural and built environment in coastal Louisiana) and determine how the national economy would respond to their loss or degradation (e.g., what is the capacity for similar components in other regions to compensate for their loss and on what time scales?). **If greater emphasis is to be placed on the national economic benefits of restoring and protecting coastal Louisiana, future planning efforts should incorporate meaningful measures of the economic benefits of these projects to the nation consistent with procedures normally employed to determine the value of a project or a suite of projects proposed as National Economic Development.**

Conversely, the significance of the coastal Louisiana wetlands to the nation in terms of both their inherent uniqueness and the ecosystem services they provide is more thoroughly documented in the LCA Study, its predecessor reports, and the scientific literature. **Although efforts to restore and protect Louisiana's wetlands will likely provide some unknown but potentially significant protection against coastal storms and hurricanes, those efforts should not be evaluated primarily on their significance for National Economic Development.**

It would appear that a more effective means to provide sediment to restore areas to the southwest and southeast of New Orleans would be to develop a large-scale diversion located closer to the Head of Passes such that most of the river water is diverted, and the river below this diversion point would essentially become a "slack water" channel but would still be maintained for navigation. This would allow most of the suspended sediments to be carried eastward to shallower water areas where they could nourish the wetlands rather than being discharged into deep water and substantially lost to the system. Finer-grained sediments could be directed into wetlands on either side of the channel. Most of the coarser bedload would either be deposited in the channel or eventually be carried out by the flows to the shallower area to be transported to the west by the waves and currents to nourish the existing barrier islands and to form new barrier islands. The slack water navigation channel would require more dredging downstream of the diversion location than is presently necessary, but this is the price if a majority of the sediments are to be captured, an essential requirement to the optimal restoration of wetlands and barrier islands. Regardless of the diversion design adopted, if the present navigation channel is to be maintained and the coarse fraction captured, substantial dredging will be required. **Some consideration should be given to an alternative or companion to the planned Third Delta, such as a larger-scale diversion closer to the Gulf of Mexico that would capture and deliver greater quantities of coarse and fine sediments for wetland and barrier island development and maintenance.**

The majority of the projects proposed in the LCA Study are based on commonly accepted, sound scientific and engineering analyses. However, it is not clear that, in the aggregate, if these projects represent a scientifically sound strategy for addressing coastal erosion at the scale of the affected area. Thus, at foreseeable rates of land loss, the level of effort described by the LCA Study will likely decrease land loss only in areas adjacent to the specific proposed projects. As found in numerous USACE policy statements and recommended in past National Research Council reports, planning and implementation of water resources projects (including those involving environmental restoration) should be undertaken within the context of the larger system (National Research Council, 2004a). This philosophy reflects the recognition that a group of projects within a given watershed or coastal system may interact at a variety of scales to produce beneficial or deleterious effects. Cost and benefit analyses discussed in the LCA Study and in supporting documents reflect an effort to identify least-cost alternatives but do not appear to reflect a system-wide effort to maximize beneficial synergies among various projects with regard to habitat loss. **The selection of any suite of individual projects in future efforts to restore coastal Louisiana should include a clear effort to maximize the beneficial, synergistic effects of individual projects to minimize or reverse future land loss.** For example, in addition to the series of relatively modest diversions and barrier island restoration efforts discussed in the LCA Study, more robust efforts may be contemplated. Two large projects not in the LCA Study, the Third Delta and abandoning the Birdsfoot Delta, may deserve greater consideration. Chapter 7 notes the gaps in knowledge that impact project choices and implementation strategies.

Finally, coastal Louisiana lies at the nexus between the Gulf of Mexico and the nation's largest watershed (the Mississippi River Basin). The current loss of wetlands and other environmental problems on and along the delta have many causes, but several of them are the result of the current management of the Mississippi River Basin. **Taking a system-wide approach to determining contributing causes and potential approaches to reducing their adverse impact on the environmental quality of coastal Louisiana should include consideration of (1) changes in the sediment flux from the basin resulting from past dam construction on the tributaries to the Mississippi River, (2) the effects of armoring the river banks, (3) the loss of wetlands in the upper part of the watershed, and (4) the impacts of runoff from agricultural operations and other activities in the Mississippi River Basin.**

7

Critical Knowledge Gaps

HIGHLIGHTS

This chapter
- Discusses proposed causal factors of wetland loss and their relevance to coastal Louisiana
- Discusses knowledge gaps associated with approaches outlined in the *Louisiana Coastal Area (LCA), Louisiana—Ecosystem Restoration Study* (LCA Study)
- Provides, where appropriate, recommendations to narrow these knowledge gaps
- Discusses stakeholder acceptance of the more aggressive possible projects
- Describes a conceptual alternative to the Third Delta that could reduce adverse stakeholder response

---

It is appropriate in an overall project at this stage and of this magnitude, such as the LCA Study (U.S. Army Corps of Engineers, 2004a), to identify critical knowledge gaps and to examine whether the actions proposed will address those gaps, whether data now exist that could reduce knowledge gaps, or whether plans should be developed to resolve these gaps. Prior to examining the knowledge gaps in detail, it is relevant to note that it might have been possible to resolve or reduce some of these knowledge gaps with available data and information. The monitoring results from the Coastal Wetlands Planning, Protection, and Restoration Act (CWPPRA) projects provide an extremely rich and unique source of information, and although considerable analysis of the monitoring data has been carried out, it is believed that significantly more general information could have been extracted and contributed to confidence in the LCA Study and design. The critical knowledge gaps lie in the causal factors of land loss—ecological, hydrological, socioeconomic, and anthropogenic (e.g., engineering). Where possible, views of the most suitable approaches to reducing these gaps are shared.

The U.S. Army Corps of Engineers' [USACE] LCA Study describes these knowledge gaps as "uncertainties" and lists the following four types (see Chapter 6):

• Type 1: Physical, chemical, geological, and biological baseline conditions
• Type 2: Engineering concepts and operational methods
• Type 3: Ecological processes, analytical tools, and ecosystem response
• Type 4: Socioeconomic and political conditions and responses

These are broad characterizations of uncertainties, and in this chapter, more specific knowledge gaps are identified. The LCA Study appropriately considers adaptive management as one approach to dealing with uncertainties. Real estate issues, considered here as one of the socioeconomic knowledge gaps, are discussed in the LCA Study.

WETLAND LOSS CAUSAL FACTORS AND RATES

It is interesting that various investigators who have studied wetland loss in Louisiana for decades hold different views as to the dominant cause(s). Possible causes include canals cut for access to oil and gas facilities; oil and gas exploration; the grazing by fur-bearing animals (i.e., nutria); and the maintenance of a fixed water course for the Mississippi River and the associated losses of freshwater, nutrients, and sediments to deep water. Although it is understood that all of these are contributors, there was a surprising divergence of assessments by experts as discussed below. Establishing the relative importance of various causes of land loss would have been helpful to the architects of the overall restoration development and execution. If the relative causes were known area by area, it would be possible to target the most appropriate solutions more effectively. The gap in knowledge of land loss rates is discussed in Box 7.1.

Box 7.1
Uncertainty in Land Loss Rates

The apparent decrease in land loss rates is not completely understood. The land loss rates peaked in the mid-1960s at more than 103 square kilometers (km^2) per yr (40 square miles [mi^2] per yr), declined to 61.9 km^2 per yr (23.9 mi^2 per yr) from 1990 to 2000, and are projected at 26.7 km^2 per yr (10.3 mi^2 per yr) from 2000 to 2050, which represents a 75 percent reduction from the mid-1960s (Figure 7.1). While some of the rates are affected by the measurement methods and explanations for some of the real reductions have been advanced, the remaining differences are so great that there appears to be a fundamental lack of understanding about the relative role of various processes in land loss. If these reductions were to continue, the prospects for no net land loss will improve substantially in the future. Thus, achieving a robust understanding of the causes of loss and changes in the rate of loss will be an important step toward determining the long-term prospects of maintaining the coastal Louisiana ecosystem and the activities it supports.

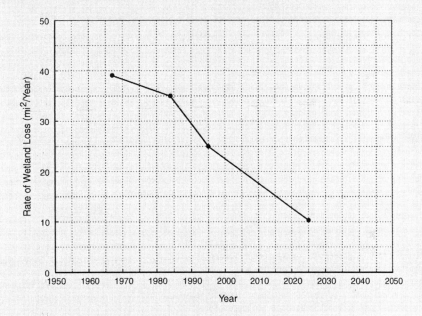

FIGURE 7.1 Measured (1956–2000) and projected (2000–2050) rates of wetland loss (data from U.S. Army Corps of Engineers, 2004a). The points are plotted at the mid-range dates of the periods represented by the wetland loss rates. (Note the rapid decrease over the last several decades.)

Dredging Versus Losses of Sediment

There is some gap in knowledge regarding the need to address the role of canal and pipeline dredging on overall land loss (Turner, 1997; Day et al., 2000). Day et al. (2000) found that based on a population of quadrangles across the state there were significant wetland losses that were not attributable to canals while agreeing with Turner (1997) that canals are an important agent in causing wetland loss in coastal Louisiana. In the Barataria and Breton Sound Basins, for example, Day et al. (2000) found that direct canal loss accounted for 47 and 68 percent of the variation in "other land loss" in a population of quadrangles. In other words, when added to the direct losses, canal effects were probably responsible for most of the wetland loss in those areas, while in rapidly subsiding areas deprived of sediment renourishment (Birdsfoot Delta) or accreting regions (Atchafalaya Basin) canal density was relatively unimportant in land loss.

In portions of the delta that are not experiencing extremely high localized subsidence, canals and pipelines with associated spoil piles are an important cause of land loss. Specific restoration projects attempting to maintain these areas will need to explore alternative methods for distributing freshwater and sediment in large quantities to more distant locations or will need to acknowledge that these are locations that will ultimately become abandoned. The need for sediment will be even more acute for restoration efforts that include large-scale plans to block damaging canals, which is a sediment-consuming task. At the same time, steps will need to be made to resolve conflicts over ownership and use of the canals and compensation for their closure. Clearly, this becomes more than a simple engineering task and a more challenging target than diversions adjacent to the Mississippi River.

Causes of Growth Fault Activity

As much as 43 percent of marsh loss in southern Louisiana is associated with erosional "hot spots." These interior marsh losses are associated with extreme rates of land subsidence (greater than 20 millimeters [mm] per yr [0.8 inches {in} per yr] and more than 10 times the worldwide average) that are an order of magnitude greater than long-term, area-wide rates of sinking (Morton et al., 2003b). There is a close spatial and temporal association between production of fluids from oil and gas fields and subsidence (as measured by both historical leveling surveys and tide gauges), and fluid withdrawal may reactivate growth faults, which lead to rapid subsidence (Morton et al., 2002, 2003b). However, there is some evidence that growth faults are a natural component of the delta environment and have been active for millions of years (Gagliano et al., 2003).

Whether water or oil and gas withdrawal accelerates land subsidence is an important question because this mechanism is not considered in modeling exercises that predict future land loss (Twilley, 2003). Morton et al. (2002) observe that as petroleum production has declined, so has the rate of land loss, suggesting there might be a future lessening of land loss independent of efforts to restore wetlands. The role of growth fault reactivation in land loss must be understood to the degree possible to better interpret the past pattern of land loss, to predict future land loss rates and locations, and to design strategies for wetland creation on the delta.

ENGINEERING KNOWLEDGE GAPS

Performance of Barrier Island Construction and Maintenance

The active processes affecting sediment transport and dispersal in the delta region are much more complex than in other areas where more experience is available on the performance of constructed barrier islands and their maintenance. These processes include severely reduced sediment supply; barriers composed of a combination of muds, silts, and sands; high rates of relative sea level rise; and episodic severe storms. Because of the complexities of these processes and the relatively limited experience related to the level of barrier island construction and maintenance required, the predictability of the performance of these constructed systems is limited.

The best avenue for developing a basic understanding of the processes attendant on the construction and evolution of barrier islands is to (1) carry out construction and maintenance on a reasonably large scale and (2) conduct comprehensive monitoring sufficient to identify pathways of sediment transport and the role of both large-scale subsidence and subsidence induced by the barrier island sediments on compressible underlying material. Barataria Basin barrier shoreline restoration was selected as one of the five main projects, and monitoring and analysis are expected to improve design capabilities substantially. Thus, the Barataria Basin barrier shoreline restoration, which includes Caminada Headland and Shell Island, is an appropriate selection for one of the five major projects.

Possibility of Accessing Richer Sediment Concentrations for Diversion

A concern in the delivery of sediments to deficient areas is the overfreshening of receiving waters, displacing desirable marine and brackish habitats such that the biota are adversely affected. One approach would be to access deeper Mississippi River waters where the sediment

Box 7.2
Sediment Delivery Quantities by the
Mississippi and Atchafalaya Rivers

There is agreement that the sediment delivery quantities in the Mississippi River system have decreased due to impoundments on the tributaries of the Mississippi River and land-use practices. The issue here is the average quantities of sediments delivered to the coast by the two rivers. The quantities of river-borne sediments that can be diverted are directly related to the area of wetlands that can be sustained and created. Table 7.1 summarizes available information from several sources. It is seen that there is at least a five-fold difference in the estimates. However, depending on the basis for the calculations used to estimate quantities required for wetland stabilization (Chapter 3), there may be adequate sediments for wetland restoration if a significant fraction of this sediment can be captured and utilized to support restoration efforts. At a minimum, this suggests that a better estimate needs to be obtained, possibly utilizing new technologies, to better define the amount of sediment available and how much sediment can realistically and economically be distributed.

concentrations should be higher; however, the quantity of sediment may not be adequate (Box 7.2). The use of siphons or pumps to deliver this slurry could perhaps provide greater sediment quantities with limited freshening not possible with near-surface intakes, which deliver less sediment per unit of water volume.

Identifying and Developing Restoration Methods
with Low-Energy Requirements

It is evident that energy costs and related constraints will continue to increase in the future. Thus, at the earliest possible stage of LCA Study development, a fundamental principle should be incorporated to select methods that will utilize the minimum amounts of energy to accomplish

TABLE 7.1 Various Estimates of Sediment Transport by the Mississippi River System

Source	Estimate of Annual Metric Tons (mt) or Cubic Meters (m^3)	Suspended, Bed Total Load (mt or m^3, plus additional comments)
Coleman, 1988	621 million mt	This estimate referred to as "sediment discharge of Mississippi" is repeated in Coleman et al. (1998).
Kesel, 1988	82 million mt (Mississippi River); 49 million mt (Atchafalaya River)	Suspended sediment for 1963–1982
Kesel et al., 1992	270 million m^3 (suspended) or 483 million mt;[a] 132 million m^3 (bedload) or 227 million mt[a]	Suspended and bedload for 1850–1895
U.S. Geological Survey, 2005	159 million mt	Mississippi River "long-term suspended"
U.S. Geological Survey, 2005	88 million mt	Atchafalaya River "sediment load"

[a]Values obtained by the following conversion factor: 1 m^3 = 1.79 mt.

the necessary water, nutrient, and sediment delivery (Box 7.3). Decisions made early in LCA Study project implementation could have far-reaching implications with the possibility of essentially binding future practice to energy-intensive methods unless expensive modifications in equipment and methodology are made in the future.

HYDROLOGIC KNOWLEDGE GAPS

The hydrologic processes can be divided into chronic and episodic knowledge gaps. The chronic knowledge gaps are the issues and processes that develop gradually over time and can be foreseen and accounted for within the Adaptive Environmental Assessment and Management Program. Potential chronic events are the gradual reduction in flows or sedi-

---

Box 7.3
Sediment Delivery Over Long Distances by Slurry Pipeline

The U.S. Army Corps of Engineers (2004a) addresses knowledge gaps of sediment delivery over long distances, a process that will be required if many of the areas of wetland degradation are to be provided with sediment. This issue could have been addressed through more effective interaction with dredging industry representatives. Slurry transport is well developed with several books written on the subject (Wasp et al., 1977; Turner, 1996; Wilson et al., 1997). Many products are delivered by pipeline slurry, including the longest distance of 1,667 kilometers (km) (1,036 miles [mi]), which conveys coal slurry from Wyoming to the southeastern states (the Energy Transportation Systems, Inc. pipeline that began operating in 1979). Although sand has a greater density than coal, and thus the slurry transport relationships differ, heavier minerals than sand (including iron and copper ore) are transported long distances in this manner.

---

ment supply to an area that alters salinities or rates of relative sea level rise. This is partially related to sediment but also to flow quantity. Knowledge gaps include the following:

- How much flow can be diverted from the Mississippi River before navigation or water supply is impacted directly or indirectly? Is this a limiting factor as more projects are introduced to the Louisiana restoration process in coming decades?
- As the small projects proposed in the LCA Study are implemented and larger ones are implemented in the future, how will the salinity regime change, and how will this affect the regional ecosystem and rates of land loss or progradation?

It is also unclear how the new conveyance channels will perform, specifically larger channels such as those planned in conjunction with the Third Delta. Will these channels be self-maintaining, and can their potential to cause damaging floods be predicted and controlled?

As demonstrated by Hurricane Katrina, episodic events, such as extreme weather, can have catastrophic consequences. The goal of Louisiana's restoration should be to minimize risk while acknowledging that similar events are inevitable. Major hurricanes, such as Katrina, and extreme river floods as experienced in 1993 result in river capture and crevasse splay features. Some aspects have been studied, such as the benefits associated with the barrier islands or the effect of hurricanes on floating marshes. As part of an effort to more fully explore the potential costs and benefits of components of Louisiana restoration and protection, questions to be addressed should include the following:

• During a future catastrophic event, could a massive investment of public restoration funds be lost in a single event? If so, how could these risks be minimized?

• How can risk to communities be minimized by the restoration program? In which areas is the risk becoming so severe and so unsustainable that managed retreat should be considered? (Since land loss will occur, sensitive subjects such as these will have to be part of the decision-making process.)

• Should local areas be temporarily or permanently evacuated until the new landscape trends become certain? (On the larger-scale diversions, the power and unpredictability of the Mississippi River are substantial, and local flood-proofing or protection measures are unlikely to be sustainable in the long term.)

There may be insufficient sediment available to economically restore all regions or even to arrest the current rate of coastal retreat within economic limitations. In the case of surface diversions, it is conceivable that a disproportionate amount of water compared to the sediment load could be diverted, resulting in sedimentation farther downstream that adversely affects the navigation channels. Another issue is how much of the current sediment discharge to the Gulf of Mexico can realistically be transferred to the coastal wetlands. To resolve these fundamental questions, detailed knowledge is required about the sediment delivery to coastal Louisiana (what proportion is derived from channel migration and bed scour or from transport from the upper Mississippi River); the relationship between bedload, suspended, and washload transport; and the factors affecting the vertical sediment concentration profile. These issues are complex and have been identified in Louisiana (Boesch et al., 1994; U.S. Army Corps of Engineers, 2004a) and in other coastal areas of the United States (Krone, 1985; Mehta and Cushman, 1989; Schoelhamer, 1996).

Several highly regarded researchers from all over the world have worked on different aspects of these questions; yet, conflicting findings were expressed during this study and in the literature. Different analytical approaches and different data sets have led to divergent conclusions. Questions that remain unanswered relate to the sediment transport characteristics of the Mississippi and Atchafalaya Rivers, optimal depositional marsh and wetland environments, and threshold force (waves and currents) levels for marsh sustainability.

When these fundamental questions have been resolved, the factors that are determined to be significant should be incorporated into a regional sediment budget, possibly based on the U.S. Geological Survey desktop model. This will allow lowest-order estimates of what combinations of diversion projects are likely to be achieved at the regional scale. This information will also be essential for use in the deterministic models.

WETLAND FORMATION KNOWLEDGE GAPS

The most effective approach to delivering siliciclastic (inorganic) sediments to accomplish optimal wetland construction is an area of uncertainty, although much knowledge has been gained through the CWPPRA projects and monitoring of the Atchafalaya–Wax Lake bayhead delta complex. Due to the lack of well-established methods for optimal wetland construction, the volume of clastic sediment delivery to areas where wetlands could be formed could be considered as a measure of LCA Study success. This recognizes the inherent time lag between sediment delivery and natural wetland development. During the initial period of the LCA Study, monitoring and interpretation of projects of various scales will improve the understanding and design capabilities of subsequent projects.

The LCA Study has identified a second critical issue that will require additional research—*in situ* production and retention of organic material in the marsh soils. The organic content of coastal marshes is often 40 percent or higher (Krone, 1985); yet, in some regions of marshes created from dredge material, the rates of organic decomposition are higher, and the buildup of the organic component is slow and uncertain (Streever, 2001). A report by Swarzenski and Doyle (2005) indicates that low soil redox potential and high sulfide levels in marshes receiving high riverine nutrient loads were more prone to organic matter decomposition and subsidence than marshes without riverine inputs. This suggests that there are limits to the capacity of marshes to assimilate nutrients and that freshwater diversions should be examined in terms of their potential for enhancing wetland loss or reducing nutrient loads to the Gulf of Mexico. These issues could have a major impact on the design considerations of created wetlands affecting the spatial area that can be restored and on the persistence of wetlands over time, as well as the amount of excess nutrients that can effectively be removed.

SOCIETAL KNOWLEDGE GAPS

Projections With and Without Project

The State of Louisiana has initiated research directed at understanding the physical, chemical, and ecological processes in coastal Louisiana.

Hydrologic and linked hydrologic-ecological models are being developed to understand the impacts of future strategies, such as those proposed under the LCA Study. Thus, resources are being used to advance the scientific understanding of these systems.

A major gap in knowledge is that the socioeconomic implications of the with- and without-project conditions are not well understood. There is little understanding of how and at what rate economic activities and population location patterns will change in the future if the coast continues to deteriorate. How and where will people be likely to leave the coast, and where will they be likely to remain? What changes will drive local economies and settlements? How will successful coastal projects alter those conditions? Thus, an ecological–socioeconomic model similar to the science models is missing. Coastal resettlement and evacuation are the "third rail" in any considerations of coastal restoration. Rather than be hidden behind the scenes, these questions should be brought forward and considered in any evaluation of proposed projects and projected populations modeled and coupled to ecosystem conditions. The social response of residents to Hurricanes Katrina and Rita should be monitored thoroughly as part of ongoing restoration efforts. Having been displaced, some residents may choose to not return, thereby reducing barriers to the initial flooding of areas to provide a means for transporting sediment to support restoration or to the strategic abandonment of other areas.

More generally, there is insufficient analysis of proposed enhancements in ecological services provided by proposed coastal restoration projects in the LCA Study or its supporting documents. The selection of projects is only marginally based on such services (e.g., storm protection) and ecological criteria. Hence, these expensive projects are not likely to provide the highest value. If resources are limited, how can they best be used to restore the coast? Should some areas of the coast go unrestored, and should resources be directed to other areas that are likely to provide more services? What will Louisiana's strategy be in dealing with the areas that cannot reasonably be protected?

Stakeholder Response

An area of substantial uncertainty is that of the near- and long-term response, reaction, and acceptance of the LCA Study and its extensions by various affected stakeholders. This issue was incorporated in the LCA Study's Type 4 uncertainty (see Chapter 6). However, the LCA Study did not adequately address this future uncertainty (perhaps beyond technical or economic knowledge gaps), which has the potential to result in cost escalations and delays that could limit the program to only moderate achievements. This was recognized implicitly in the LCA Study's near-

Box 7.4
Sediment Delivery and Stakeholder Concerns

The first example addresses the delivery of sediments over broad areas of coastal subsidence. Many subsiding areas are not privately owned nor are they being used for agricultural or other productive purposes. Stakeholder acceptance of programs to deliver sediments to these areas is anticipated. However, much of the subsiding area is privately owned and used in some form of production ranging from residential areas to agriculture to the harvesting of marine resources. Levees have been constructed to prevent flooding of many agricultural areas. With subsidence, such areas will become less productive and more expensive to farm until eventually a decision is made by the owner to abandon them. If this decision could be made by all owners in a general area at the same time, it would be feasible to consider delivery of water and sediment by some means to reinstate or maintain elevations compared to relative sea level rise. However, even with this concession, the ongoing subsidence would require either a near-continuous delivery of sediment to the area or an initial delivery that would increase the land elevation by several decimeters. This would be followed by usage of the land productively, if possible, and several years or decades later, a repetition of the sediment delivery process. This mimics the natural processes. The newly deposited sediment would require several years to achieve a near-optimum productivity level. The above scenario is fraught with stakeholder difficulties. It is unlikely that individual landowners would all be convinced to participate in this program, and litigation and delays in land acquisition for this purpose would increase the time and cost of such a program. At some stage, real estate costs could increase to the point that restoration would not be economically viable in some areas. Conversely, the value of areas prone to frequent inundation may fall.

term development by the selection of projects that would encounter relatively limited stakeholder resistance. Two examples illustrating the concerns are provided below (Boxes 7.4 and 7.5).

ECOLOGICAL KNOWLEDGE GAPS

Value of Habitat Shifts

Habitat switching model predictions (U.S. Army Corps of Engineers, 2004a) indicate that restoration activities will shift much of the wetland habitat to fresher regimes (fresh and intermediate marsh) with significant

---

Box 7.5
The Third Delta and Stakeholder Concerns

The second example of stakeholder concern is the possible LCA Study centerpiece—the Third Delta. Current preliminary plans are to consider the formation of the Third Delta, commencing with a diversion in the vicinity of Donaldsonville and a 88.5-km (55-mi) long new conveyance channel to discharge points within the Barataria and Terrebonne Bays. This channel to some degree would mimic and parallel the natural Bayou Lafourche system. Two magnitudes of *maximum* peak flow are considered when the Mississippi River floods: 3,398 cubic meters (m^3) per sec (4,444 cubic yards [yd^3] per sec) and 6,796 m^3 per sec (8,889 yd^3 per sec). The historic Bayou Lafourche carried 15 percent of the total flow of the Mississippi River in the 1850s, at a time when the Atchafalaya River carried 12 percent of the total flow (Kesel, 2003). Although there is no direct basis for conversion, the average flows associated with the maximum of the two flow magnitudes considered (6,796 m^3 per sec [8,889 yd^3 per sec]) would be substantially less than the amount carried historically, and the sediment delivery would be even less due to the long-term decrease in sediment discharge, arguing for a more aggressive diversion effort. Although the proposed designs would capture fine sediment, a very small portion of the coarser bedload component essential for the maintenance of barrier islands would be diverted. Either of these two design levels would continue to divert most of the coarse load into deep water through the Birdsfoot Delta or would require interception and capturing upstream of the exits from the Birdsfoot Delta. In these designs, the potential real estate issues, including costs and litigation, of a 88.5-km (55-mi) long channel commencing near Donaldsonville through, and possibly flooding many stakeholders properties, would seem to be difficult if not a near impossibility.

---

loss of brackish and sometimes saline marsh. This will favor specific ecosystem components at the expense of others. For example, habitat for adult croaker, menhaden, spotted seatrout, and oysters will decline in some regions, while habitat and productivity of mink, dabbling ducks, and alligator will increase. The ecological and economic benefits of these shifts are unclear and require further attention.

Because of poor data quality, most of the predicted habitat shifts are based on models with large uncertainties. It is imperative to identify parameters that are most critical to model output and variability (via sensitivity analysis) and to identify the level of uncertainty for the data currently applied to these parameters since they will drive decision making.

Nutrient Trapping by Wetlands and Effect on the
Gulf of Mexico "Dead Zone"

While there is still considerable uncertainty about the role of nutrient input from the Mississippi River watershed in controlling the extent and intensity of hypoxia in the Gulf of Mexico (Rabalais and Turner, 2001; Rabalais et al., 2002b), it is generally accepted that reduction of these inputs will be beneficial. There is little information about the value of different wetland habitats in absorbing nutrients or depositing organics bound to particles. More vegetation is clearly better than less because it slows the flow of water, thereby increasing nutrient and sediment residence time in wetlands prior to their reaching the ocean and increasing the probability of uptake or deposition. To confirm amelioration of hypoxia as a result of the proposed coastal Louisiana restoration effort, it will be necessary to quantify the anticipated nutrient reductions, to identify the mechanisms and their relationship to different wetland habitats, and to assess the consequences to the open ocean system. The anticipated reduction in nutrient inputs to the Gulf of Mexico are likely to be no more than 10 percent (Mitsch et al., 2001) and could be substantially less if most nutrients enter the wetland during high flow and flood conditions.

ADDRESSING GAPS IN THE EXISTING KNOWLEDGE BASE

After reviewing the LCA Study, key engineering, hydrologic, wetland formation, societal, and ecological knowledge gaps have been identified that, if addressed, will improve the likelihood for restoration success. **Explicit steps to address these gaps should be incorporated in the Science and Technology Program called for in the LCA Study.** These gaps include the following:

- The relative importance of various causes of land loss by area to more effectively target the solutions
- The causes of loss and changes in the rate of loss to determine the long-term prospects of maintaining the coastal Louisiana ecosystem and the activities it supports
- The role of growth fault reactivation in land loss to interpret past land loss patterns, predict future land loss rates and locations, and design strategies for land creation on the delta
- The potential for various methods of sediment delivery over long distances (perhaps using existing dredging techniques), which impacts cost and feasibility of projects
- The relationship between bedload, suspended, and washload transport to understand the factors affecting the vertical sediment concentration profile

- The factors to incorporate into a regional sediment budget to be able to use deterministic models effectively
- The economic and societal toll of land loss used to justify, partially, the restoration of coastal Louisiana
- Stakeholders' near- and long-term responses to gauge their acceptance of the restoration activities
- The ecological and economic benefits of habitat shifts to better inform the project selection process
- Quantification of the anticipated nutrient reduction, identification of the mechanisms and their relationship to different wetland habitats, and assessment of the consequences to the open ocean system to assess amelioration of hypoxia
- Quantification of the anticipated risk reduction from hurricanes due to storm surge and wave activity

8

Findings and Recommendations

The *Louisiana Coastal Area (LCA), Louisiana—Ecosystem Restoration Study* (LCA Study) is the product of many years of planning, scientific investigation, field evaluation of methods for wetland restoration, and consensus building. Those responsible for the development to this stage are commended for their success in overcoming difficult and broad obstacles. However, some elements of the LCA Study raise serious concerns, which are summarized later in this chapter.

Land loss in coastal Louisiana is occurring at a rate that will increase risk to lives and property from storms, as demonstrated by Hurricane Katrina, and will harm ecological, industrial, and agricultural systems in the region. The challenge of protecting and restoring this wetland system is unprecedented in geographic scope, in the variety of the causative forces and the variable role each plays at any given time or location, and in the diversity and intensity of competing stakeholder interests.

Land losses in coastal Louisiana are caused by a variety of natural processes and human actions. However, the role of each causal factor varies with location and time, and the relative contribution of each at a given location is not well quantified. More or less ubiquitous causes of land loss include reduced sediment load in the river due to dams and levees throughout the river basin, reduced sedimentation in the delta due to channelization of the lowermost reaches of the Mississippi River by levees (which direct much of the river-borne sediments to deep water at the terminus of the Birdsfoot Delta), grazing by fur-bearing animals (e.g., nutria), and the natural wetland systems' tendency to erode in some areas and build in others. Superimposed on these broad influences are relative

161

sea level rise and localized land loss "hot spots" occurring along growth faults, access canals, and navigation waterways. Some of the individual causative factors encompass both natural and anthropogenic elements. For example, the extraction of oil and gas near growth faults may speed local subsidence events. Further, public interests often support conflicting management policies. For example, maintenance of the canal system along the lowermost reaches of the Mississippi River benefits the shipping industry, but levees and spoil banks along their margins limit delivery of freshwater, sediments, and nutrients to areas in need of wetland restoration and maintenance. Eliminating levees heightens risk of flooding to businesses and homes located along canals and rivers. Restoring the essential and widespread distribution of sediment and freshwater flow, while maintaining stakeholder acceptance of the adverse impacts of such efforts, will be the overarching challenge to future comprehensive efforts to realize the vision espoused by *Coast 2050: Toward a Sustainable Coastal Louisiana* (Coast 2050).

SOUNDNESS OF APPROACH AND PERFORMANCE METRICS

In general, the strategies employed in the LCA Study are based on methodologies that have been demonstrated and proven largely through the nearly 15 years of efforts carried out under the Coastal Wetland Planning, Protection, and Restoration Act (CWPPRA). The LCA Study differs from early efforts, such as CWPPRA, in that (1) it recognizes and attempts to address the need for understanding the natural and social systems that are shaping coastal Louisiana today, and (2) it identifies specific actions over the next 5–10 years to address land loss at specific locations in the area. The budget proposed in the LCA Study gives the two components roughly equal weight. This reflects a strong commitment to developing longer-term efforts to restore and protect Louisiana.

Establishing Realistic Expectations

As discussed in some length in Chapters 2 and 3, full restoration of past Louisiana wetland cover and function will not be possible. The natural and anthropogenic processes contributing to net land loss in coastal Louisiana are significant and pervasive, and they have been operating for decades. Achieving no net loss is not a feasible objective because the social, political, and economic impediments are extensive; the sediment supply is limited; and the affected area is large. The LCA Study's five restoration features would reduce land loss by about 20 percent from 26.7 square kilometers (km^2) per yr (10.3 square miles [mi^2] per yr) to 22.3 km^2 per yr (8.6 mi^2 per yr). More extensive wetland protection than proposed in the

LCA Study would, obviously, require greater efforts to reduce land loss at much greater expense. These facts have to be broadly appreciated to avoid widespread disappointment with the LCA projects.

Louisiana's coastal restoration plans must acknowledge these limitations prominently and adjust goals and public expectations accordingly. Given that even under optimistic assumptions, the Louisiana coast will continue to suffer land loss; therefore, the emphasis should be on establishing realistic estimates of future landforms and conveying these to stakeholders. **Restoration efforts should be focused to maximize targeted ecological, social, and economic benefits while promoting managed retreat in selected regions.** (Since land loss will occur in selected areas, sensitive subjects such as these will need to be part of the decision-making process.) This could involve reducing the rate of land loss in key areas and allowing the system to approach natural equilibrium in others. Future efforts must focus more realistically on the location patterns of human settlements relative to project locations, including the option of infrastructure depreciation and abandonment.

Further, since there is a finite availability of water flow and sediment and most of the restoration activities will take decades to provide maximum results, care should be taken to ensure that implementation of an individual project will not preclude other strategies or elements in the future. **To achieve this, the development of an explicit map of the expected future landscape of coastal Louisiana should be a priority as the implementation of the LCA Study moves ahead.** Such an explicit declaration of the proposed "end state" of restoration efforts in Louisiana provides an important performance metric. Development of such a map will also require meaningful stakeholder involvement and the commitment of decision makers at all levels of local, state, and federal governments.

Local Land-Use Planning and Zoning

As discussed in Chapter 3, efforts to restore significant portions of coastal Louisiana would entail changing the current geographic distribution of land, water, and wetland. Land use and infrastructure development (e.g., roads, pipelines, utilities) have changed in response to the changing coastline. The proposed projects will again force change in the way people work, live, and play in the area. One way to deal efficiently with the change is through comprehensive land-use planning that is coordinated with the planned restoration projects.

A survey of the local parish governments reveals that 10 of the 20 parish governments have a comprehensive plan, and at least four plans are more than 10 years old. All but one of the parishes has a planning department; one is in the Department of Public Works and two others are citizen commissions. All have subdivision regulations and floodplain management and require building permits so there is familiarity with the idea of government regulation of land. **The parishes should develop comprehensive land-use plans in order for there to be orderly and economically efficient relocation of infrastructure, homes, and businesses during coastal restoration (as planned for in the LCA Study).** The cost of developing such plans could be part of the cost of the projects borne by the State of Louisiana (National Research Council, 2000c). Clearly, effective land-use plans that act in concert with and support a comprehensive restoration effort will require a widely understood and accepted "end state" of restoration efforts.

Taking a Systems Approach to Coastal Restoration in Louisiana

As discussed above and in Chapters 5 and 6, most of the individual projects proposed in the LCA Study are based on commonly accepted, sound scientific and engineering analyses. However, it is not clear that, in the aggregate, if these projects represent a scientifically sound strategy for addressing coastal erosion at the scale of the affected area. Thus, at foreseeable rates of land loss, the level of effort described by the LCA Study will likely decrease land loss only in areas adjacent to the specific proposed projects. As found in numerous U.S. Army Corps of Engineers (USACE) policy statements and recommended in past National Research Council reports, planning and implementation of water resources projects (including those involving environmental restoration) should be undertaken within the context of the larger system (National Research Council, 2004a). This philosophy reflects the recognition that a group of projects within a given watershed or coastal system may interact at a variety of scales to produce beneficial or deleterious effects. Cost-effectiveness analyses discussed in the LCA Study and in supporting documents reflect an effort to identify least-cost alternatives but do not appear to reflect a system-wide effort to maximize beneficial synergies among various projects with regard to habitat loss. **The selection of any suite of individual projects in future efforts to restore coastal Louisiana should include a clear effort to maximize the beneficial, synergistic effects of individual projects to minimize or reverse future land loss.**

Coastal Louisiana lies at the nexus between the Gulf of Mexico and the nation's largest watershed (the Mississippi River Basin). The current loss of lands and other environmental problems on and along the delta have many causes, but several of them are the result of the current management of the Mississippi River Basin. **Taking a system-wide approach to determining contributing causes and potential approaches to reducing their adverse impact on the environmental quality of coastal Louisi-**

ana should include consideration of (1) changes in the sediment flux from the basin resulting from past dam construction on the tributaries to the Mississippi River, (2) the effects of armoring the river banks, (3) the loss of lands in the upper part of the watershed, and (4) the impacts of runoff from activities within the Mississippi River Basin.

Large-Scale Delivery Systems

Annual land loss rates in coastal Louisiana have varied over the last 50 years, declining from a maximum of 100 km^2 per yr (39 mi^2 per yr) for the period 1956–1978 to as little as 64 km^2 per yr (25 mi^2 per yr) in the 1990s (U.S. Army Corps of Engineers, 2004a). Although land loss rates appear to have fallen, perhaps reflecting that the loss of the most vulnerable wetlands has already occurred, rates still remain high (26.7 km^2 per yr [10.3 mi^2 per yr]). Furthermore, actual land building will be experienced only in areas adjacent to the implemented projects.

As discussed in Chapter 6, keeping pace with the relative subsidence rate over the entire delta would require the delivery of sediment-laden waters over long distances, would be extremely costly, and would have an adverse impact on a significant subset of stakeholders. Even accomplishing the objective of maximum wetland restoration and maintenance with reasonable expenditures within strategically selected portions of coastal Louisiana will require much larger-scale projects than are proposed in the LCA Study or have been carried out under CWPPRA. Carrying out these larger-scale projects will likely increase the number of stakeholders who experience adverse impacts from the projects themselves. Although adverse impacts cannot be fully eliminated, preference should clearly be given to larger-scale projects that provide maximum benefit in terms of net land gain, while minimizing the adverse impacts on stakeholders and reducing overall costs per unit of land gained.

Though the size of the area it would impact would still make it controversial, some consideration should be given to an alternative or companion to the planned Third Delta, such as a larger-scale diversion closer to the Gulf of Mexico that would capture and deliver greater quantities of coarse and fine sediments for wetland and barrier island development and maintenance. This diversion could occur above Head of Passes and capture most of the river discharge such that the present channel to the south would essentially be a "slack water" channel that would continue to serve navigation. This would allow a substantial portion of the suspended sediment and a portion of the coarse sediment to nourish the wetlands and barrier islands. A substantial portion of the remaining coarse sediment fraction would be deposited in the slack water channel and would require dredging for navigational purposes. This dredged material could then be placed predominantly to the west where it would further nourish, maintain, and form barrier islands under the westward-directed natural dispersion. The underlying objective of these efforts should be to capture and utilize significant portions of both the fine and the coarse sediment loads of the Mississippi River.

ADDRESSING KNOWLEDGE GAPS

Chapter 7 discusses a number of technical issues that represent significant knowledge gaps, including the following:

- The causes of loss and changes in the rate of loss to determine the long-term prospects of maintaining the coastal Louisiana ecosystem and the activities it supports
- The geographic variability in the relative role played by natural and anthropogenic causes of land loss to more effectively target the appropriate solutions
- The role of growth fault reactivation in land loss to interpret past land loss patterns, to predict future land loss rates and locations, and to design strategies for land creation on the delta
- The limited understanding of the feasibility of engineered methods of sediment delivery over long distances (such as those employed by the dredging industry) to evaluate the cost and feasibility of projects that do not rely on natural processes to distribute sediment
- The relationship between bedload, suspended, and washload transport to understand the factors affecting the vertical sediment concentration profile
- A robust regional sediment budget to use deterministic models effectively
- The economic and societal toll of land loss to frame restoration of coastal Louisiana in terms of national relevance
- Stakeholders' near- and long-term responses to gauge their acceptance of the restoration activities

Overall, these knowledge gaps define two overarching concerns regarding the future magnitude of efforts needed to offset land loss and the acceptance by stakeholders of various large-scale projects. Both of these concerns will best be addressed by proceeding with the LCA Study (with some modifications as discussed throughout this report), while anticipating and responding to information as it becomes available. The Science and Technology (S&T), demonstration project, and adaptive management programs will contribute to the reduction of these and other knowledge gaps through exploring areas of present uncertainty, collecting and analyzing the monitoring data, and where appropriate, providing input to the charting of new directions and methodologies. There is need, however, to consider some modifications to the S&T, demonstration project, and adaptive management programs, how they interact, and how they are integrated into the overall program management scheme.

Program Management

As discussed in Chapters 3, 4, and 5, effective management of the efforts proposed in the LCA Study will be a critical factor leading to the overall success of the restoration effort in Louisiana. Expanded efforts described in this report will place further burden on the management structure. The management plan, as described in the LCA Study, lacks clarity concerning the institutional mechanisms proposed in the LCA Study that will be used for decision making and accountability. The proposed decision support system will contribute to resolving this deficiency, and this system should be subject to external peer review. The stakeholder component, which appears to have been effective in the CWPPRA program, is not represented adequately in the efforts described in the LCA Study.

Successful implementation of the LCA Study's restoration strategies will depend in part on how well the program is being managed. These strategies rely heavily on the interaction of numerous groups and committees that provide direction, assessment, and feedback to the program through an adaptive management process. This is a difficult challenge for the program in and of itself and is only further complicated by state and federal agencies, having to work in full harmony with each other.

The Adaptive Environmental Assessment and Management process for the LCA Study is an appropriate effort to integrate emerging technical information into the management process, in addition to the use of sound science in understanding the efficacy of past actions in order to modify or change future actions. **Steps should be taken to strengthen the Adaptive Environmental Assessment and Management process throughout the management structure.**

CWPPRA, Coast 2050, and efforts to develop *Louisiana Coastal Area, LA—Ecosystem Restoration: Comprehensive Coastwide Ecosystem Restoration Study* (draft LCA Comprehensive Study) placed greater emphasis on stakeholder involvement than is proposed in the LCA Study. **Stakeholder participation (including, at a minimum, representation from local government, industry, key organizations, and citizens at large) should to be accounted for in the management structure of the Louisiana coastal area program.**

Model Development and Application

As discussed in Chapter 5, the LCA Study proposes the development of process-based models for prediction of coastal response as a central feature of current and future restoration efforts. Modeling will be a key component of the design, operation, and maintenance of the restoration and management of coastal Louisiana. It will also be a valuable tool in evaluating the concerns of stakeholders and explaining the efforts and potential outcomes to diverse audiences because it will enable the consequences of various management alternatives and the regional effects to be more fully understood.

To achieve this objective, it is important that the models used are defensible, accessible, and transparent so that they can be used with confidence, and the uncertainties in model outputs can be quantified. **The model codes employed should reflect widely accepted and verified approaches with a community-wide effort at model development and maintenance. The models should also utilize open-source codes with an active program of model refinement that includes quality control, consistent data sets by all users, and appropriately available and useful data. The management of data, tracking of model data sets, calibration of model parameters, and interaction and coordination of model users and developers are important aspects that should be included in the management plan.** This effort should be structured to attract synergistic collaborations among modelers worldwide and enhance the current extensive regional expertise in federal agencies, state agencies, and academia.

Integrating Emerging Information into Management Decisions

The knowledge gaps discussed in Chapter 7 can be grouped into two major categories: (1) understanding spatial and temporal trends in future land loss rates without the LCA Study's proposed projects and (2) stakeholder response. These gaps can be reduced through careful monitoring and program implementation when a meaningful adaptive management strategy is employed. Thus, the adaptive management program will play a major role in collecting and synthesizing data and charting new directions as appropriate. **The S&T Program requires a more explicit statement of program responsibilities and means for setting priorities; it must be integrated more effectively into the central management structure through the adaptive management process and include better representation of social sciences and ecological processes.** Additional key questions relate to the major causes of land loss, recognizing that the relative role of various processes is location dependent. The future rates of loss are uncertain, and some evidence suggests that the average rate of land loss across coastal Louisiana may be decreasing. Documented rates of worldwide sea level rise and regional subsidence clearly indicate that in the absence of adequate action, land loss in coastal Louisiana will continue. If, however, rates of land loss are indeed declining, the potential to more fully offset land loss may be greater.

The S&T Program envisioned in the LCA Study is an innovative and essential element and provides a process for planning and assimilating monitoring results and developing adaptive management strategies. In addition, the S&T Program is an appropriate administrative home for model development and maintenance. The proposed S&T Program represents a very positive step in the development of a process to address the need for improved understanding of how coastal Louisiana may respond to various restoration efforts or may evolve in the absence of some of those efforts. However, it is unreasonable to expect any region to have all of the necessary experience and human resources to address most effectively the challenges of the magnitude represented by land loss in coastal Louisiana. Just as the funding of the LCA Study and its extensions includes a combination of state and federal resources, the scientific and other elements of the LCA Study should draw on the best state, national, and international talents available. Therefore, **the LCA Study should direct efforts toward capacity building that enables the program to address its stated objectives by drawing on the widest possible pool of national and international technical expertise.**

UNDERSTANDING COSTS AND BENEFITS

Although the resources at risk (e.g., industry, agriculture, fisheries, ecosystems, urban areas, petroleum, traditional cultures of the delta) were discussed and, where appropriate, their magnitudes described, the LCA Study included no concerted effort to establish a quantitative link between LCA projects and benefits of storm damage reduction, fisheries improvement, or land building. As understanding of the short- and long-term economic impacts of Katrina and Rita becomes clearer, a more meaningful effort to evaluate the national economic significance of protecting the natural and built environment in coastal Louisiana will be possible. While such information will provide an important context for decision making, it will still be crucial to understand the role wetlands play in protecting specific components of the overall system and to determine how specific restoration efforts can enhance that protection.

ECONOMIC JUSTIFICATION

The LCA Study states that "execution of the LCA [Study] would make significant progress towards achieving and sustaining a coastal ecosystem that can support and protect the environment, economy, and culture of southern Louisiana and, thus, contribute to the economy and well-being of the nation" (U.S. Army Corps of Engineers, 2004a). The economic analysis provided within the LCA Study and its supporting documents, however, evaluates alternative approaches to meet stated ecosystem restoration objectives, as is consistent with USACE policy for evaluating projects proposed as National Environmental Restoration efforts. Evaluating the benefits to the nation of restoring coastal Louisiana, as implied by the statement of task, would have required USACE planners to carry out cost-benefit analyses more consistent with a National Economic Development appraisal. USACE officials appeared to view the efforts described within the LCA Study as following under National Environmental Restoration as opposed to National Economic Development appraisal. Consequently, they did not attempt to identify and meaningfully quantify the contribution to the economy of the nation. Lacking this information, the committee could not determine whether the economic benefits to the nation are large or small relative to the costs of implementing the LCA projects. Such an analysis would require significant effort and resources that were beyond those available to the committee in the nine months following the release of the LCA Study in November 2004. This said, some characteristics of such an analysis can be articulated.

As discussed in Chapter 6, the LCA Study presents sufficient information about the importance of some components of the natural and built environment in coastal Louisiana (e.g., system of deep water ports, oil and gas receiving and transmission facilities, complex and extensive urban landscape, robust commercial fishery) to suggest that substantial economic interests are at stake in coastal Louisiana and that these interests have national significance. The immediate impacts of Katrina underscore the importance of New Orleans, and adjacent areas of the Gulf Coast, in the national economy. Establishing the true, national economic significance of efforts to restore coastal wetlands in Louisiana as proposed in the LCA Study, however, must go beyond simply identifying and characterizing these components and should include an analysis of how specific restoration efforts will preserve or enhance the value of these components (i.e., some restoration efforts may have little influence on the vulnerabilities of specific components of natural and built environment in coastal Louisiana) and determine how the national economy would respond to their loss or degradation (e.g., what is the capacity for similar components in other regions to compensate for their loss and on what time scales?). If, as implied by the statement of task, greater emphasis is to be placed on the national economic benefits of restoring and protecting coastal Louisiana, future planning efforts should incorporate meaningful measures of the economic significance of these projects to the nation consistent with procedures normally employed to determine the value of a project or a suite of projects for National Economic Development.

As greater understanding is gained of the short- and long-term economic impacts of Katrina, a more meaningful effort to evaluate the national economic significance of protecting the natural and built environment in coastal Louisiana will be possible. Such information would provide an important context for decision making; however, it will still be important to understand the role that wetlands play in protecting specific components of the overall system and to determine how specific restoration efforts can enhance this protection. Although wetlands and adjacent barrier islands and levees are known to reduce impacts from waves, their role in reducing storm surge is complex and less predictable. Surges contain multiple components, including barometric tide effects, wind stress-induced setup, wave-induced setup, and Coriolis forces. As was pointed out repeatedly in the public media during Katrina and Rita, in the northern hemisphere the eastern side of a hurricane tends to drive water northward in a counterclockwise manner. If a storm stalls off a coast for a significant period of time, it will continue to drive water onshore for a prolonged period, regardless of the nature of any intervening wetland or barrier island. Thus, the potential for reducing risk due to storm surge is more difficult to generalize.

Conversely, the significance of the coastal Louisiana wetlands to the nation in terms of both their inherent uniqueness and the ecosystem services they provide is more thoroughly documented in the LCA Study, its predecessor reports, and the scientific literature. **Although efforts to restore and protect Louisiana's wetlands will likely provide some unknown but potentially significant protection against coastal storms and hurricanes, those efforts should not be evaluated primarily on their significance for National Economic Development.**

Although many of the projects considered during development of the LCA Study are larger than many funded under CWPPRA, for the most part, they are still significantly smaller than the large-scale diversions discussed in Chapter 6 and 7. Thus, with some exceptions, project selection in the LCA Study is based on the comparative analysis of a large number of potentially small projects. The end result of evaluating mostly small projects is that mostly small projects emerged as recommendations. On the other hand, the broader wetland ecosystem needs protection from marine processes at the seaward margins of the system and increased inputs of sediments and freshwater from landward. Addressing problems of this scale with a series of small projects is likely to make little difference unless the projects are designed to maximize synergistic effects or are coupled with some larger-scale efforts. Small projects may be useful for learning how the system works and for developing the confidence necessary for larger-scale project selection. However, it is unlikely that these small projects, operating more or less independently, will have a significant, positive impact on problems of the magnitude currently experienced in coastal Louisiana. Hence, it would seem unlikely that the economic benefits of the projects specifically proposed by the LCA Study (as opposed to wetlands restoration in general) will be large relative to the program costs.

DEVELOPING A COMPREHENSIVE PLAN

Many of the shortcomings of the LCA Study identified in this report may be attributed to the rather brief amount of time (less than one year) taken to produce the near-term plan presented in the LCA Study from the draft LCA Comprehensive Study. Much of the stakeholder involvement and economic analysis that supported the restoration framework review that characterized the draft comprehensive plan was absent (in the case of stakeholder involvement) or significantly weakened (in the case of restoration feature selection) in the LCA Study. (While a comprehensive plan is needed, this does not necessarily imply endorsement of *the* draft LCA Comprehensive Study.)

The phased approach that characterizes the LCA Study has advantages and disadvantages. The main advantage is a definite time over which to implement and evaluate specific projects, revise the projections of benefits from program extensions, and plan these extensions. Both the executors and the supporters of the LCA Study will, therefore, be conscious of the need to produce significant results and demonstrate success. In this regard, the phased approach will have a salutary effect. The disadvantages of the phased approach in the LCA Study include the same need to demonstrate solid progress in the 10-year project period. The sorting criteria related to timing have resulted in the selection of projects that were already in the USACE and CWPPRA planning pipeline. The formalized temporal constraint on projects selected precluded consideration of projects with solid potential for long-term benefits that had not yet been fully designed (and, thus, cannot be undertaken in five years or completed in 10).

Given this temporal constraint, it is important to note that, by definition, the activities proposed within the LCA Study are intended to lay a foundation for more effective and robust efforts to preserve and protect coastal Louisiana. As discussed above, the LCA Study points out that implementing the restoration efforts proposed would reduce this land loss by about 20 percent. Furthermore, actual land building will be experienced only in areas adjacent to the implemented projects. The significant investment represented by these projects and the efforts to develop the tools and understanding necessary to support future restoration and protection efforts will yield a substantial return of benefits only if future projects are carried out in a comprehensive manner. The funding needed to carry out the activities described in the LCA Study should be recognized as the first of many increments that will be required if substantial progress is to be made.

Project Selection Methodology

As discussed in Chapters 5 and 6, in addition to lacking fundamental transparency, the project selection process as documented in the LCA Study creates a somewhat inaccurate appearance of legitimacy and rigor. There is insufficient attention to the large knowledge gaps surrounding project benefits measurement, including the somewhat arbitrary weighting of various ecological and physical endpoints of projects. The project selection process primarily uses ecological benefits early on in project formulation then uses least-cost alternatives for aggregates of projects (referred to as "frameworks" by USACE) as a filtering criterion to accept and reject frameworks based on their socioeconomic value. However, since the physical and ecological relationships between projects in a framework are not clear, and frameworks optimized for cost include many projects that are not chosen for implementation, the actual role of socioeconomic factors in project selection is not clear.

Furthermore, although the cost-effectiveness analysis was carried out on frameworks, the selection decision was made for individual features. Since the cost-effectiveness was calculated for groups, there appears to be some potential for individual features that might score poorly if singled out during a cost-effectiveness analysis to be elevated by more cost-effective projects in the same group. Since the selection process then breaks the groups down into individual features, it would seem more appropriate to consider the cost-effectiveness of individual features, unless the projects can be shown to be physically, ecologically, or logistically interrelated. The rationale for this analysis is poorly articulated in the LCA Study, reinforcing the need for greater transparency. Obviously, if the more comprehensive approach called for in this report were used, determining the cost-effectiveness of a single project in the absence of all others would not be appropriate.

These criteria and the need to demonstrate solid near-term success likely resulted in the avoidance of bold innovative projects that would

(1) affect significant sediment delivery to the system, such as abandonment of the Birdsfoot Delta; (2) maximize synergistic effects for reducing land loss over longer time scales by the selection of strategically located or larger-scale projects; or (3) address some of the difficult issues associated with stakeholder response. While the efforts preceding the LCA Study have achieved a laudable degree of unanimity among stakeholders on the conceptual restoration plan, this unanimity will be tested by the difficult decisions associated with implementation of the larger-scale projects needed to achieve greater sediment, water, and nutrient delivery over a larger area more effectively. **The project selection procedure requires more explicit accounting of the synergistic effects of various projects and improved transparency of project selection to sustain stakeholder support. Furthermore, beneficial, synergistic interaction among projects cannot be assumed but should be demonstrated through preconstruction analysis.**

It is important to note that by definition, the activities proposed within the LCA Study are intended to lay a foundation for more effective and robust efforts to preserve and protect coastal Louisiana. By its own analysis, the LCA Study points out that constructing the five restoration features it proposes would reduce land loss by about 20 percent (from 26.7 km^2 per yr [10.3 mi^2 per yr] to 22.3 km^2 per yr [8.6 mi^2 per yr]) at an estimated total cost of $864 million (or $39,400 per hectare [$15,900 per acre]) over the 50-year life of the projects, not including maintenance and operational costs. Furthermore, actual land building will be experienced only in areas adjacent to the implemented projects. The significant investment represented by these projects and the efforts to develop the tools and understanding necessary to support future restoration and protection efforts will yield a substantial return of benefits only if future projects are carried out in a comprehensive manner. The funding required to carry out the activities described in the LCA Study should be recognized as the first of many increments that will be required if substantial progress is to be made. **A comprehensive plan to produce a more clearly articulated future distribution of land in coastal Louisiana is needed. Such a plan should identify clearly defined milestones to be achieved through a series of synergistic projects at a variety of scales.** (While a comprehensive plan is needed, this does not necessarily imply endorsement of the draft LCA Comprehensive Study, which was not formally released by USACE or reviewed as part of this study.) The review detailed in this report found no instance where the proposed activities, if initiated, would preclude development and implementation of a more comprehensive approach. Conversely, many examples were identified where implementing the proposed activities would support a more comprehensive approach. **Thus, the efforts proposed in the LCA Study should be implemented, except** where specific recommendations for change have been made in this report and only in conjunction with the development of a comprehensive plan.

The Aftermath of Katrina and Rita

As the State of Louisiana and the nation begin to recover from Katrina and Rita, efforts to restore wetlands in Louisiana will undoubtedly compete with reconstruction and levee maintenance or enhancement efforts. As this report and numerous other National Research Council reports have pointed out, efforts to design and implement water resource projects (including environmental restoration and flood control projects) should be carried out within a watershed and coastal system context. **Efforts to rebuild the Gulf Coast and reduce coastal hazards in the area, therefore, should be integral components of an effective and comprehensive strategy to restore and protect coastal Louisiana wetlands.**

References

Antweiler, R.C., D.A. Goolsby, and H.E. Taylor. 1995. Nutrients in the Mississippi River. In *Contaminants in the Mississippi River, 1987-1992*, Meade, R.H. (ed.). U.S. Geological Survey Circular 1133, Reston, VA.

Barras, J., S. Beville, D. Britsch, S. Hartley, S. Hawes, J. Johnston, P. Kemp, Q. Kinler, A. Martucci, J. Porthouse, D. Reed, K. Roy, S. Sapkota, and J. Suhayda. 2003. *Historic and Predicted Coastal Louisiana Land Changes: 1978-2050*. U.S. Geological Survey Open File Report 03-334. U.S. Geological Survey, National Wetlands Research Center, Baton Rouge, LA.

Benjamin, D.K. 2001. Common law and environmental protection. *PERC Reports* 19(3):12.

Bentley, S.J., K. Rotondo, H.H. Roberts, G.W. Stone, and O.K. Huh. 2003. Transport and accumulation of cohesive fluvial sediments on the inner shelf: From the Atchafalaya River downdrift to the Louisiana Chenier Plain. *Proceedings of the International Conference on Coastal Sediments 2003*. World Scientific Publishing Corporation and East Meets West Productions, Corpus Christi, TX.

Biedenharn, D.S. 1995. *Lower Mississippi River Channel Response: Past, Present and Future*. Ph.D. Dissertation, Department of Civil Engineering, Colorado State University, Fort Collins, CO.

Biedenharn, D.S., C.R. Thorne, and C.C. Watson. 2000. Recent morphological evolution of the lower Mississippi River. *Geomorphology* 34:227-249.

Blumberg, A.F. and G.L. Mellor. 1987. A description of a three-dimensional coastal model. In *Three Dimensional Coastal Ocean Models*, Heaps, N.S. (ed.). American Geophysical Union, Washington, DC.

Boesch, D.F., M.N. Josselyn, A.J. Mehta, J.T. Morris, W.K. Nuttle, C.A. Simenstad, and D.J.P. Swift. 1994. Scientific assessment of coastal wetland loss, restoration, and management in Louisiana. *Journal of Coastal Research Special Issue* 20:1-103.

Brooks, M. 2002. *Planning Theory for Practitioners*. Planners Press, American Planning Association, Chicago, IL.

Caffey, R.H. and B. Leblanc. 2002. Closing the Mississippi River Gulf Outlet: Environmental and economic considerations. In *Interpretive Topic Series on Coastal Wetland Restoration in Louisiana*, Coastal Wetland Planning, Protection, and Restoration Acts Outreach Committee (eds.). National Sea Grant Library No. LSU-G-02-004. Baton Rouge, LA.

Coleman, J.M. 1988. Dynamic changes and processes in the Mississippi River Delta. *Geological Society of America Bulletin* 100:999-1015.

Coleman, J.M. and H.H. Roberts. 1988a. Late Quaternary depositional framework of the Louisiana continental shelf and upper continental slope. *Transactions Gulf Coast Association of Geological Societies* 38:407-419.

Coleman, J.M. and H.H. Roberts. 1988b. Sedimentary development of the Louisiana continental shelf related to sea level cycles: Part I—Sedimentary sequences. *Geo-Marine Letters* 8:63-108.

Coleman, J.M., H.H. Roberts, and G.W. Stone. 1998. Mississippi River Delta: An overview. *Journal of Coastal Research* 14(3):698-716.

Costanza, R., F.H. Sklar, and J.W. Day. 1987. Using the Coastal Ecological Landscape Spatial Simulation (CELSS) model for wetland management: Coastal Zone '87. *Proceedings of the 5th Symposium on Coastal and Ocean Management* 4:3879-3890.

Costanza, R., F.H. Sklar, M.L. White, and J.W. Day. 1989. Modeling landscape dynamics in the Atchafalaya/Terrebonne marshes of coastal Louisiana using the CELSS model: Executive Summary. In *Final Report to U.S. Fish and Wildlife Service*. U.S. Fish and Wildlife Service, Washington, DC.

Davis, D.W. 2000. Historical perspective on crevasses, levees, and the Mississippi River. In *Transforming New Orleans and Its Environs: Centuries of Changes*, Colten, C.E. (ed.). University of Pittsburgh Press, PA.

Davis, M.S. 2002. *Legal Issues Associated with Coastal Restoration*. Presentation to the National Research Council, December 12, 2002, New Orleans, LA.

Day, J.W., G.P. Shaffer, L.D. Britsch, D.J. Reed, S.R. Hawes, and D. Cahoon. 2000. Pattern and process of land loss in the Mississippi Delta: A spatial and temporal analysis of wetland habitat change. *Estuaries* 23(4):425-438.

DeLaune, R.D., R.H. Baumann, and W.H. Patrick. 1983. Relationships among vertical accretion, coastal submergence, and erosion in a Louisiana Gulf Coast marsh. *Journal of Sedimentary Petrology* 53:147-157.

DeLaune, R.D., J.A. Nyman, and W.H. Patrick. 1994. Peat collapse, ponding, and wetland loss in a rapidly subsiding coastal marsh. *Journal of Coastal Research* 10:1021-1030.

Draut, A.E., G.C. Kineke, D.W. Velasco, M.A. Allison, and R.J. Prime. 2005. Influence of the Atchafalaya River on recent evolution of the Chenier Plain inner continental shelf, northern Gulf of Mexico. *Continental Shelf Research* 25:91-112.

Environmental Protection Agency. 2005. *Mississippi River Basin and Gulf of Mexico Hypoxia: Culture/History*. [Online]. Available: http://www.epa.gov/msbasin/culture.htm [2005, October 11].

Gagliano, S.M. 1994. *An Environmental Economic Blueprint for Restoring the Louisiana Coastal Zone: The State Plan*. Report of the Governor's Office of Coastal Activities, Science Advisory Panel Workshop. Coastal Environments, Inc., Baton Rouge, LA.

Gagliano, S.M. and J.L. Van Beek. 1975. An approach to multiuse management in the Mississippi delta system. In *Deltas: Models for Exploration*, Broussard, M.S. (ed.). Houston Geological Society, TX.

Gagliano, S.M., E.B. Kemp, K.M. Wicker, and K.S. Wiltenmuth. 2003. *Active Geological Faults and Land Change in Southeastern Louisiana*. Prepared for the U.S. Army Corps of Engineers. Coastal Environments, Inc., Baton Rouge, LA.

Gould, H.R. and E. McFarlan. 1959. Geologic history of the Chenier Plain, southwestern Louisiana. *Transactions, Gulf Coast Association of Geological Societies* 9:261-270.

Harmar, O.P. 2004. *Morphological and Process Dynamics of the Lower Mississippi River.* Ph.D. Dissertation, University of Nottingham, UK.

Howe, H.V., R.J. Russell, B.C. McGuirt, and B.C. Craft. 1935. Report on the geology of Cameron and Vermillion Parishes. In *Geological Bulletin No. 6.* Louisiana Geological Survey.

Hoyt, J.H. 1969. Chenier versus barrier: Genetic and stratigraphic distinction. *American Association of Petroleum Geologists Bulletin* 53:299-306.

Huh, O.K., N.D. Walker, and C. Moeller. 2001. Sedimentation along the eastern Chenier Plain coast: Down drift impact of a delta complex shift. *Journal of Coastal Research* 17: 72-81.

Innes, J.E. 1995. Planning theory's emerging paradigm: Communicative action and interactive practice. *Journal of Planning Education and Research* 16(3):183-198.

Innes, J.E. 1998. Information in communicative planning. *Journal of the American Planning Association* 64(1):52-63.

Keown, M.P., E.A. Dardeau, Jr., and E.M. Causey. 1981. *Characterization of the Suspended-Sediment Regime and Bedload Gradation of the Mississippi River Basin.* Volumes 1 and 2, Report 1, U.S. Army Corps of Engineers, Vicksburg, MS.

Kesel, R.H. 1988. The decline in suspended load of the lower Mississippi River and its influence on adjacent wetlands. *Environmental Geology and Water Sciences* 11:271-281.

Kesel, R.H. 1989. The role of the Mississippi River in wetland loss in southeastern Louisiana, USA. *Environmental Geology and Water Sciences* 13(3):183-193.

Kesel, R.H. 2003. Human modification of the sediment regime of the lower Mississippi River flood plain. *Geomorphology* 56:325-334.

Kesel, R.H., E.G. Yodis, and D.J. McGraw. 1992. An approximation of the sediment budget of the lower Mississippi River prior to major human modification. *Earth Surface Processes and Landforms* 17:711-722.

Kolb, C.R. and J.R. Van Lopik. 1966. Depositional environments of Mississippi River deltaic plain, southeastern Louisiana. In *Deltas in their Geological Framework,* Shirley, M.L. (ed.). Houston Geological Society, Houston, TX.

Krone, R.B. 1985. Simulation of marsh growth under rising sea levels. *Hydraulics and Hydrology in the Small Computer Age,* Waldrop, W.R. (ed.). Proceedings of the Specialty Conference Sponsored by the Hydraulics Division of the American Society of Civil Engineers.

Lake Pontchartrain Basin Foundation. 1995. *Comprehensive Management Plan for the Pontchartrain Basin.* [Online]. Available: http://www.saveourlake.org/comprehensive. html [2005, June 6].

Lake Pontchartrain Basin Foundation. 2005. *The Mississippi River Gulf Outlet.* [Online]. Available: http://www.saveourlake.org/mississippi.html [2005, June 6].

Louisiana Coastal Wetlands Conservation and Restoration Task Force. 1993. *Coastal Wetlands Planning, Protection and Restoration Act 3rd Priority Project List Report.* U.S. Army Corps of Engineers, U.S. Department of Agriculture, U.S. Department of Commerce, U.S. Department of the Interior, Environmental Protection Agency, and Louisiana Department of Natural Resources. Louisiana Department of Natural Resources, Baton Rouge, LA. 215 pp.

Louisiana Coastal Wetlands Conservation and Restoration Task Force. 2003. *The 2003 Evaluation Report to the U.S. Congress on the Effectiveness of Coastal Wetland Planning, Protection and Restoration Act Projects.* Louisiana Department of Natural Resources, Baton Rouge, LA.

Louisiana Coastal Wetlands Conservation and Restoration Task Force and the Wetlands Conservation and Restoration Authority. 1998. *Coast 2050: Toward a Sustainable Coastal Louisiana.* Louisiana Department of Natural Resources, Baton Rouge, LA.

Louisiana Mid-Continent Oil and Gas Association. 2003. *Louisiana Oil and Gas Parish Profile—2003*. [Online]. Available: http://www.lmoga.com/pchart.html [2005, October 5].

Martin, J.F., M.L. White, E. Reyes, G.P. Kemp, H. Mashriqui, and J.W. Day. 2000. Evaluation of coastal management plans with a spatial model: Mississippi Delta, Louisiana, USA. *Environmental Management* 26:117-129

Martin, L.R. 2002. *Regional Sediment Management: Background and Overview of Initial Implementation*. IWR Report 02-PS-2. U.S. Army Corps of Engineers, Institute for Water Resources, Alexandria, VA.

Meade, R.H. 1995. Setting: Geology, hydrology, sediments and engineering of the Mississippi River. In *Contaminants in the Mississippi River, 1987-1992*, Meade, R.H. (ed.). U.S. Geological Survey Circular 1133, Reston, VA.

Meade, R.H. and R.S. Parker. 1985. Sediment in rivers of the United States. *National Water Summary 1984—Water Quality Issues*. U.S. Geological Survey.

Mehta, A.J. and R.M. Cushman (eds). 1989. *Workshop on Sea Level Rise and Coastal Processes*. Publication number DOE/NBB-0086. U.S. Department of Energy, Washington, DC.

Minerals Management Service. 2005. *Section 8(g) Disbursements to States, FY 1986-2000*. [Online]. Available: http://www.mrm.mms.gov/stats/pdfdocs/disboff.pdf [2005, July 27].

Mississippi River/Gulf of Mexico Watershed Nutrient Task Force. 2001. *Action Plan for Reducing, Mitigating, and Controlling Hypoxia in the Northern Gulf of Mexico*. Washington, DC.

Mitsch, W.J., J.W. Day, W. Gilliam, P.M. Groffman, D.L. Hey, G.L. Randall, and N. Wang. 2001. Reducing nitrogen loading to the Gulf of Mexico from the Mississippi River Basin: Strategies to counter a persistent ecological problem. *Bioscience* 51:373-388.

Morton, R.A., N.A. Buster, and M.D. Krohn. 2002. Subsurface controls on historical subsidence rates and associated wetland loss in southcentral Louisiana. *Transactions, Gulf Coast Association of Geological Societies* 52:767-778.

Morton, R.A., G. Tiling, and N.F. Ferina. 2003a. *Primary Causes of Wetland Loss at Madison Bay, Terrebonne Parish, Louisiana*. U.S. Geological Survey Open-File Report 03-60. U.S. Geological Survey, Center for Coastal and Watershed Studies, St. Petersburg, FL.

Morton, R.A., G. Tiling, and N.F. Ferina. 2003b. Causes of hot-spot wetland loss in the Mississippi Delta plain. *Environmental Geosciences* 10:71-80.

National Marine Fisheries Service. 2003. *Fisheries of the United States, 2002*. Fisheries Statistics and Economics Division, Silver Spring, MD.

National Research Council. 1999a. *New Strategies for America's Watersheds*. National Academy Press, Washington, DC.

National Research Council. 1999b. *New Directions in Water Resources Planning for the U.S. Army Corps of Engineers*. National Academy Press, Washington, DC.

National Research Council. 2000a. *Clean Coastal Waters: Understanding and Reducing the Effects of Nutrient Pollution*. National Academy Press, Washington, DC.

National Research Council. 2000b. *Reconciling Observations of Global Temperature Change*. National Academy Press, Washington, DC.

National Research Council. 2000c. *Watershed Management for Potable Water Supply: Assessing the New York City Strategy*. National Academy Press, Washington, DC.

National Research Council. 2002. *Review Procedures for Water Resources Project Planning*. National Academy Press, Washington, DC.

National Research Council. 2004a. *River Basins and Coastal Systems Planning Within the U.S. Army Corps of Engineers*. The National Academies Press, Washington, DC.

National Research Council. 2004b. *Adaptive Management for Water Resources Project Planning*. The National Academies Press, Washington, DC.

Neill, C. and L.A. Deegan. 1986. The effect of Mississippi River Delta lobe development on the habitat composition and diversity of Louisiana coastal wetlands. *American Midland Naturalist* 116:296-303.

Office of Administrative Law Judges Law Library. 1991. *Dictionary of Occupational Titles: Glossary*. [Online]. Available: http://www.oalj.dol.gov/public/dot/refrnc/glossary. htm [2005, June 6].

Orth, K., J.W. Day, D.F. Boesch, E.J. Clairain, W.J. Mitsch, L. Shabman, C. Simenstad, B. Streever, C. Watson, J. Wells, and D. Whigham. 2005. Lessons learned: An assessment of the effectiveness of a National Technical Review Committee for oversight of the plan for the restoration of the Mississippi Delta. *Ecological Engineering* 25:153-167.

Ozawa, C. and E. Seltzer. 1999. Taking our bearings: Mapping a relationship among planning practices, theory, and education. *Journal of Planning Education and Research* 18(3): 257-266.

Penland, S., R. Boyd, D. Nummendal, and H.H. Roberts. 1981. Deltaic barrier development of the Louisiana coast. *Supplemental Transactions, Gulf Coast Association of Geological Societies* 31:471-465.

Penland, S. and J.R. Suter. 1989. Geomorphology of the Mississippi River Chenier Plain. *Marine Geology* 90:231-58.

Penland, S., P.F. Connor, Jr., and A. Beall. 2004. Changes in Louisiana's shoreline: 1855–2002. In *Louisiana Coastal Area (LCA), Louisiana—Ecosystem Restoration Study: Appendix D— Louisiana Gulf Shoreline Restoration Report*. U.S. Army Corps of Engineers, New Orleans District, LA.

Rabalais, N.N. and R.E. Turner. 2001. Hypoxia in the northern Gulf of Mexico: Description, causes, and change. In *Coastal Hypoxia: Consequences for Living Resources and Ecosystems*, Rabalais, N.N. and R.E. Turner (eds.). Coastal and Estuarine Studies, American Geophysical Union, Washington, DC.

Rabalais, N.N., R.E. Turner, and D. Scavia. 2002a. Beyond science into policy: Gulf of Mexico hypoxia and the Mississippi River. *Bioscience* 52:1291-1242.

Rabalais, N.N, R.E. Turner, Q. Dortch, D. Justic, V.J. Bierman, Jr., and W.J. Wiseman, Jr. 2002b. Nutrient-enhanced productivity in the northern Gulf of Mexico: Past, present, and future. *Hydrobiologia* 475/476:39-63.

Reading, H.G. (ed.). 1978. Deltas. In *Sedimentary Environments and Facies*. Elsevier, New York, NY.

Reed, D.J. 1995. Sediment dynamics, deposition, and erosion in temperate salt marshes. *Journal of Coastal Research* 11:295.

Reyes, E., J.F. Martin, M.L. White, J.W. Day, and G.P. Kemp. 2003. Habitat changes in the Mississippi Delta: Future scenarios and alternatives. In *Explicit Landscape Modeling*, Voinov, A. and R. Costanza, (eds.). Springer-Verlag, New York, NY.

Rittel, H.W.J. and M.M. Webber. 1973. Dilemmas in a general theory of planning. *Policy Sciences* 4:155-169.

Robbins, L.G. 1977. *Suspended Sediment and Bed Material Studies on the Lower Mississippi River*. Potamology Investigation Report 300-1. U.S. Army Corps of Engineers, Vicksburg, MS.

Roberts, H.H. 1997. Dynamic changes of the Holocene Mississippi River Delta plain: The delta cycle. *Journal of Coastal Research* 13:605-627.

Roberts, H.H. 1998. Delta switching: Early responses to the Atchafalaya River diversion. *Journal of Coastal Research* 14(3):882-899.

Roberts, H.H., S. Bentley, J.M. Coleman, S.A. Hsu, O.K. Huh, K. Rotondo, M. Inoue, L.J. Rouse, Jr., A. Sheremet, G. Stone, N. Walker, S. Welsh, and W.J. Wiseman, Jr. 2002. Geological framework and sedimentology of recent mud deposition on the eastern Chenier Plain coast and adjacent inner shelf, western Louisiana. *Transactions, Gulf Coast Association of Geological Societies* 52:849-859.

Schoelhamer, D.H. 1996. Factors affecting suspended-solids concentration in South San Francisco Bay, California. *Journal of Geophysical Research* 101(C5):12,087-12,095.

Scott, L.C. 2002. *The Energy Sector: Still a Giant Economic Engine for the Louisiana Economy.* [Online]. Available: http://www.lmoga.com/information/economicreport.pdf [2005, June 3].

Sheremet, A. and G.W. Stone. 2003. Observations of nearshore wave dissipation over muddy sea beds. *Journal of Geophysical Research* 108(C11):3357.

Sklar, F.H., R. Costanza, and J.W. Day. 1985. Dynamic spatial simulation modeling of coastal wetland habitat succession. *Ecological Modelling* 29:261-281.

Smith, S.R. and G.A. Jacobs. 2005. Seasonal circulation fields in the northern Gulf of Mexico calculated by assimilating current meter, shipboard ADCP, and drifter data simultaneously with the shallow water equations. *Continental Shelf Research* 25:157-183.

Snowden, J.O., W.C. Ward, and J.R.J. Studlick. 1980. *Geology of Greater New Orleans: Its Relationship to Land Subsidence and Flooding.* New Orleans Geological Survey, Inc., New Orleans, LA.

Stone, G.W., A. Sheremet, X. Zhang, Q. He, B. Liu, and B. Strong. 2003. Landfall of two tropical storms seven days apart along south central Louisiana, USA. *Proceedings of the International Conference on Coastal Sediments 2003.* World Scientific Publishing Corporation and East Meets West Productions, Corpus Christi, TX.

Streever, B. 2001. *Saving Louisiana? The Battle for Coastal Wetlands.* University Press of Mississippi, Jackson, MS.

Stronach, J.A., J.O. Backhaus, and T.S. Murty. 1993. An update on the numerical simulation of the waters between Vancouver Island and the mainland: The GF8 model. *Oceanography and Marine Biology Annual Review* 31:1-86.

Swarzenski, C.M. and T.W. Doyle. 2005. *Pore-Water and Substrate Quality of the Peat Marshes at the Barataria Preserve, Jean Lafitte National Historical Park and Preserve, and Comparison with Penchant Basin Peat Marshes, South Louisiana, 2000-2002.* U.S. Geological Survey Scientific Investigations Report 2005-5121. U.S. Geological Survey, St. Petersburg, FL.

Syvitski, J.P.M., C.J. Vörösmarty, A.J. Kettner, and P. Green. 2005. Impact of humans on the flux of terrestrial sediment to the global coastal ocean. *Science* 308(5720):376-380.

Tardo, R. 2003. *Future of MRGO Not Much Rosier than Past.* [Online]. Available: http://www.louisianasportsman.com/stories/2003/paradise-lost/future-of-mrgo.htm [2005, June 6].

Thorne, C.R., O.P. Harmar, and N. Wallerstein. 2000. *Sediment Transport in the Lower Mississippi River.* U.S. Army Corps of Engineers, Washington, DC.

Turner, R.E. 1997. Wetland loss in the Northern Gulf of Mexico: Multiple working hypotheses. *Estuaries* 20:1-13.

Turner, R.E. and N.N. Rabalais. 1991. Changes in Mississippi River water quality this century—Implications for coastal food webs. *Bioscience* 41:140-147.

Turner, R.E., and N.N. Rabalais. 1994. Coastal eutrophication near the Mississippi River Delta. *Nature* 368:619-621.

Turner, T.M. 1996. *Fundamentals of Hydraulic Dredging,* 2nd ed. American Society of Civil Engineers Press, New York, NY.

Twilley, R.R. 2003. *Conceptual Ecological Models for Planning and Evaluating the Louisiana Coastal Area Ecosystem Restoration Plan.* Presented at Botany 2003, July 26-31, 2003, Mobile, AL. Botanical Society of America, St. Louis, MO.

Uchupi, E. 1975. Physiography of the Gulf of Mexico and Caribbean Sea. In *The Ocean Basins and Margins, Vol. 3: The Gulf of Mexico and the Caribbean,* Nairn, A.E.M. and F.G. Stehli (eds.). Plenum Press, New York, NY.

U.S. Army Corps of Engineers. 1999a. *Section 905(b) (WRDA 1986) Analysis Louisiana Coastal Area, Louisiana—Ecosystem Restoration.* New Orleans District, LA.

U.S. Army Corps of Engineers. 1999b. *Policy Guidance Letter 61: Application of Watershed Perspective to Corps of Engineers Civil Works Programs and Activities*. [Online]. Available: http://www.usace.army.mil/inet/functions/cw/cecwp/branches/guidance_dev/pgls/pdf/pgl61.pdf [2005, October 6].

U.S. Army Corps of Engineers. 2001. *Rivers Project Master Plan—Section IV: Regional Description and Factors Influencing Development*. [Online]. Available: http://www.mvs.usace.army.mil/pm/Riverplan/MasterPlan.html [2005, August 17].

U.S. Army Corps of Engineers. 2002a. *Regional Sediment Management Research Program Fact Sheet*. [Online]. Available: http://chl.wes.army.mil/research/sedimentation/RSM/FactSheetRSMP.pdf [2005, October 6].

U.S. Army Corps of Engineers. 2002b. *Regional Sediment Management Research Program: Draft*. [Online]. Available: http://www.wes.army.mil/rsm/pubs/pdfs/ProgDescription RSMP.pdf [2005, October 6].

U.S. Army Corps of Engineers. 2003a. *Louisiana Coastal Area, LA—Ecosystem Restoration: Comprehensive Coastwide Ecosystem Restoration Study*. New Orleans District, LA.

U.S. Army Corps of Engineers. 2003b. *Waterborne Commerce of the United States: Calendar Year 2003*. [Online]. Available: http://www.iwr.usace.army.mil/ndc/wcsc/pdf/wcusvgc03.pdf [2005, June 3].

U.S. Army Corps of Engineers. 2003c. *U.S. Army Corps of Engineers Dredging Contracts Awarded*. [Online]. Available: http://www.iwr.usace.army.mil/ndc/dredge/pdf/awards03.pdf [2005, June 3].

U.S. Army Corps of Engineers. 2004a. *Louisiana Coastal Area (LCA), Louisiana—Ecosystem Restoration Study*. New Orleans District, LA.

U.S. Army Corps of Engineers. 2004b. *Scoping Report: Louisiana Coastal Area (LCA) Ecosystem Restoration Study*. New Orleans District, LA.

U.S. Army Corps of Engineers. 2005a. *Breaux Act (Coastal Wetlands Planning, Protection and Restoration Act)*. [Online]. Available: http://www.lacoast.gov/cwppra/slideshow/cwppra-TIcommittee-18feb05.pdf [2005, June 6].

U.S. Army Corps of Engineers. 2005b. *Chief of Engineers Report*. [Online]. Available: http://www.lca.gov/nearterm/lca_signed_chief_report_31Jan05.pdf [2005, October 19].

U.S. Department of Agriculture. 2002. *2002 Census of Agriculture—Volume 1, Chapter 2: Louisiana Parish Level Data*. [Online]. Available: http://www.nass.usda.gov/census/census02/volume1/la/index2.htm [2005, September 27].

U.S. Department of the Interior (Fish and Wildlife Service) and U.S. Department of Commerce (U.S. Census Bureau). 2003. *2001 National Survey of Fishing, Hunting, and Wildlife-Associated Recreation: Louisiana*. [Online]. Available: http://www.census.gov/prod/2003pubs/01fhw/fhw01-la.pdf [2005, September 27].

U.S. Geological Survey. 2005. *LaCoast*. [Online]. Available: http://www.lacoast.gov/ [2005, May 26].

U.S. Senate. 2002. *Complete Statement of Lieutenant General Robert B. Flowers, Chief of Engineers, U.S. Army Corps Of Engineers, Before the Committee on Environment and Public Works, United States Senate, on Water Resources Development Programs within the U.S. Army Corps of Engineers*. [Online]. Available: http://epw.senate.gov/107th/Flowers_061802.htm [2005, October 6].

Walsh, J.P., N.W. Driscoll, and C.A. Nittrouer. In press. Understanding the distal delta: Controls on the fate of fine-grained sediments in fluvial dispersal systems. *Geology*.

Wasp, E.J., J.P. Kenny, and R.L. Gandhi. 1977. *Solid-Liquid Flow: Slurry Pipeline Transport*. Trans Tech Publications, Germany.

Wells, F.C. 1980. *Hydrology and Water Quality of the Lower Mississippi River*. Louisiana Office of Public Works Technical Report 21. Department of Transportation and Development, Baton Rouge, LA.

Wells, J.T. and G.P. Kemp. 1981. Atchafalaya mud streams and recent mudflat progradation: Louisiana Chenier Plain. *Gulf Coast Association of Geological Transactions* 31:409-416.

Wilson, K.C., G.R. Addie, A. Sellgren, and R. Clift. 1997. *Slurry Transport Using Centrifugal Pumps*, 2nd ed. Blackie, Academic and Professional, London, UK.

Wiseman, W.J. and S.P. Dinnel. 1988. Shelf currents near the mouth of the Mississippi River. *Journal of Physical Oceanography* 18:1287-1291.

Wright, L.D. 1985. River deltas. In *Coastal Sedimentary Environments*, Davis, R.A. (ed.). Springer-Verlag, New York, NY.

Wright, L.D. 1995. *Morphodynamics of Inner Continental Shelves*. CRC Press, Inc., Boca Raton, FL.

Wright, L.D. and C.A. Nittrouer. 1995. Dispersal of river sediments in coastal seas: Six contrasting cases. *Estuaries* 18:494-508.

Wright, L.D., C.R. Sherwood, and R.W. Sternberg. 1997. Field measurements of fairweather bottom boundary layer processes and sediment suspension on the Louisiana inner continental shelf. *Marine Geology* 140:329-345.

Appendix
A

Committee and Staff Biographies

COMMITTEE

Robert Dean (*chair*) is a professor of civil and coastal engineering at the University of Florida. Dr. Dean earned a Sc.D. in 1959 from Massachusetts Institute of Technology. His research focuses on clarifying wave theories, beach erosion, and coastal processes by combining field measurements with relevant analytical procedures. Dr. Dean is a member of the National Academy of Engineering.

Jeffrey Benoit is director of the Coastal and Marine Program at SRA International, Inc. Mr. Benoit earned an M.S. in coastal geology in 1978 from Georgia Institute of Technology-Skidaway Institute of Oceanography. His research focuses on coastal management, marine conservation, policy analysis, program assessment, and coastal hazard mitigation planning.

Stephen Farber is a professor of economics at the University of Pittsburgh and director of the Environmental Decision Support Program and the Public and Urban Affairs Program. Dr. Farber earned a Ph.D. in economics in 1973 from Vanderbilt University. His research focuses on the economics of ecosystems, particularly the economic valuation of ecosystems and their services.

Reinhard E. Flick is an oceanographer at the California Department of Boating and Waterways and a research associate at the Scripps Institution of Oceanography's Center for Coastal Studies. Dr. Flick earned a Ph.D. in 1978 in oceanography from Scripps Institution of Oceanography. His research focuses on shoreline erosion assessment, environmental impacts of engineering works, inlet hydrology, and tidal processes.

Margot Garcia was an associate professor in the Department of Urban Studies and Planning at Virginia Commonwealth University, from which she retired in June 2003. Dr. Garcia earned a Ph.D. in watershed management in 1980 from the University of Arizona. Her research focuses on environmental planning, especially a watershed approach to water quality, bicycle planning and safety, citizen participation, and waste management.

Peter Goodwin is the DeVlieg Presidential Professor in Ecohydraulics at the University of Idaho. Dr. Goodwin earned a Ph.D. in hydraulic engineering in 1986 from the University of California, Berkeley. His research focuses on modeling physical processes in natural aquatic systems.

Daniel Huppert is a professor of marine affairs and an adjunct associate professor of economics at the University of Washington. Dr. Huppert earned a Ph.D. in economics in 1975 from the University of Washington. His research focuses on fish and wildlife planning and coastal ecosystems management.

Joseph Kelley is a professor of marine science in the Department of Geological Sciences and School of Marine Sciences at the University of Maine. Dr. Kelley earned a Ph.D. in geology in 1980 from Lehigh University. His research focuses on measuring sea level change and the response of shorelines to that change.

Lisa Levin is a professor of integrative oceanography at Scripps Institution of Oceanography. Dr. Levin earned a Ph.D. in biological oceanography from Scripps Institution of Oceanography. Her research focuses on population and community ecology of soft-sediment habitats, wetlands ecology and restoration, larval dispersal and the influence on life histories on population dynamics, ecology of deep-sea reducing environments, ecosystem-level consequences of species invasion, and animal-sediment-plant-geochemical interactions.

Scott Nixon is a professor of oceanography at the University of Rhode Island. Dr. Nixon earned a Ph.D. in botany in 1970 from the University of North Carolina. His research focuses on productivity and biogeochemical cycling of coastal ecosystems with an emphasis on estuaries, lagoons, and wetlands. Dr. Nixon is a previous member of the Ocean Studies Board.

John M. Teal is scientist emeritus at Woods Hole Oceanographic Institution. Dr. Teal earned a Ph.D. in 1955 from Harvard University. His research has focused on coastal wetlands, effects of hydrostatic pressure on deep-sea animals, physiology of warm-blooded fishes, bird migration over the oceans, oil pollution, wastewater treatment, and restoration ecology. He is currently working principally on coastal and inland wetland restoration.

L. Donelson Wright is the Chancellor Professor of Marine Science in the Virginia Institute of Marine Science at the College of William and Mary. Dr. Wright earned a Ph.D. from Louisiana State University. His research focuses on coastal and shoreface morphodynamics and benthic boundary layer processes and variability, nearshore and estuarine oceanography, waves, sediment transport processes, river-mouth deltaic and estuarine processes, and river effluent dynamics.

STAFF

Dan Walker *(study director)* earned a Ph.D. in geology in 1990 from the University of Tennessee. A scholar at the Ocean Studies Board, Dr. Walker also holds a joint appointment as a Guest Investigator at the Marine Policy Center of the Woods Hole Oceanographic Institution. Since joining the Ocean Studies Board in 1995, he has directed a number of studies including *Oil Spill Dispersants: Efficacy and Effects* (2005), *Future Needs in Deep Submergence Science: Occupied and Unoccupied Vehicles in Basic Ocean Research* (2004), *Environmental Information for Naval Warfare* (2003), *Oil in the Sea III: Inputs, Fates and Effects* (2002), *Spills of Emulsified Fuels: Risks and Response* (2002), *Clean Coastal Waters: Understanding and Reducing the Effects of Nutrient Pollution* (2000), *Science for Decisionmaking: Coastal and Marine Geology at the U.S. Geological Survey* (1999), *Global Ocean Sciences: Toward an Integrated Approach* (1998), and *The Global Ocean Observing System: Users, Benefits, and Priorities* (1997). A member of the American Geophysical Union, the Geological Society of America, and the Oceanography Society, Dr. Walker was recently named Editor of the *Marine Technology Society Journal*. A former member of both the Kentucky and North Carolina State geologic surveys, his interests focus on the value of environmental information for policy making at local, state, and national levels.

Jodi Bostrom is a research associate with the Ocean Studies Board. She earned a B.S. in zoology in 1998 from the University of Wisconsin-Madison. Since starting with the Ocean Studies Board in May 1999, Ms. Bostrom has worked on several studies pertaining to fisheries, marine mammals, nutrient overenrichment, and ocean exploration. She will earn an M.S. in environmental science from American University in December 2006.

Appendix
B

Acronyms and Abbreviations

AEAM   Adaptive Environmental Assessment and
                Management

CELSS   Coastal Ecological Landscape Spatial Simulation
                Model
cm   centimeter
Coast 2050   *Coast 2050: Toward a Sustainable Coastal Louisiana*
CWPPRA   Coastal Wetlands Planning, Protection, and
                Restoration Act

DHI   Danish Hydraulic Institute
draft LCA   *Louisiana Coastal Area, LA—Ecosystem Restoration:*
  Comprehensive   *Comprehensive Coastwide Ecosystem Restoration*
  Study   *Study*

EPA   Environmental Protection Agency

ft   foot
FWS   U.S. Fish and Wildlife Service

in   inch
IWR   Institute of Water Resources

km   kilometer
km²   square kilometer

LCA Study	*Louisiana Coastal Area (LCA), Louisiana—Ecosystem Restoration Study*
LERRD	Land, easements, rights of way, relocation, and disposal
m	meter
m^3	cubic meter
mi	mile
mi^2	square mile
mm	millimeter
MRGO	Mississippi River Gulf Outlet
mt	metric ton
NOAA	National Oceanic and Atmospheric Administration
NRC	National Research Council
NRCS	National Resources Conservation Service
NTRC	National Technical Review Committee
PEIS	programmatic environmental impact statement
PMT	program management team
POM	Princeton Ocean Model
PPL	project priority list
S&T	science and technology
sec	second
USACE	U.S. Army Corps of Engineers
USDA	U.S. Department of Agriculture
USGS	U.S. Geological Survey
WRDA	Water Resources Development Act
yd^3	cubic yard
yr	year